JavaScript Design Patterns

Deliver fast and efficient production-grade JavaScript
applications at scale

Hugo Di Francesco

JavaScript Design Patterns

Group Product Manager: Rohit Rajkumar

Publishing Product Manager: Kushal Dave

Senior Content Development Editor: Feza Shaikh

Technical Editor: Simran Udasi

Copy Editor: Safis Editing

Project Coordinator: Aishwarya Mohan

Indexer: Subalakshmi Govindhan

Production Designer: Jyoti Kadam

Marketing Coordinators: Nivedita Pandey and Anamika Singh

First published: March 2024

Production reference: 1150224

Published by Packt Publishing Ltd.

Grosvenor House

11 St Paul's Square

Birmingham

B3 1RB, UK

ISBN 978-1-80461-227-9

www.packtpub.com

To my wife, Amalia, for being my first supporter in all my endeavors.
To my daughter, Zoë, for making me want to show that the impossible sometimes is.

– Hugo Di Francesco

Contributors

About the author

Hugo Di Francesco is a software engineer who has worked extensively with JavaScript. He holds an MEng degree in mathematical computation from **University College London** (**UCL**). He has used JavaScript across the stack to create scalable and performant platforms at companies such as Canon and Elsevier, and in industries such as print on demand and mindfulness. He is currently tackling problems in the travel industry at Eurostar with Node.js, TypeScript, React, and Kubernetes, while running the eponymous Code with Hugo website. Outside of work, he is an international fencer, in the pursuit of which he trains and competes across the globe.

I want to thank all the people who have supported me in my life and writing journey, particularly my wife Amalia, and my family.

About the reviewers

Dr. Murugavel, a distinguished and versatile educator in the realms of computer science engineering and information technology. With over 13 years of enriching experience at renowned universities and an additional 8+ years dedicated to the dynamic field of data analytics, Dr. Murugavel stands as a beacon of expertise at the intersection of academia and technology.

His journey is marked by successive achievements, particularly in handling core subjects and programming languages, with a keen emphasis on practical knowledge. As a mentor and guide for major projects, Dr. Murugavel actively engages in groundbreaking research within his specialized field. His commitment to bridging theory and application has made him a valuable resource for students and researchers alike.

His technical proficiency extends across a spectrum of disciplines. He is well-versed in full stack web development, SQL, data analytics, Python, and BI tools, showcasing theoretical knowledge and a hands-on understanding of these technologies. His extensive portfolio includes the development of numerous applications using JSP, ASP, and ASP.NET, reflecting his prowess in both frontend and backend development.

In the realm of databases, he demonstrates versatility across MS-SQL Server, MySQL, MongoDB, Django, MS Access, Oracle, and FoxPro. His proficiency in various **Integrated Development Environments (IDEs)** and tools such as Anaconda, Visual Studio, GitHub, JBuilder, JCreator, MATLAB, Sublime 3, and Adobe Dreamweaver further solidifies his standing in the technological landscape.

In the realm of data science and **Business Intelligence** (**BI**) tools, his skills are extensive, encompassing PowerBI, DAX, VBA Macros for Excel, SSAS, and SSIS. His ability to harness these tools illuminates the path to insightful data analysis and visualization.

Shubham Thakur, a dynamic senior software engineer (A3 grade) at EPAM, specializes in technologies such as JavaScript, Angular, Next.js, Node, MySQL, MongoDB, AWS Cloud, and IoT. His expertise in these domains has significantly contributed to his project successes. He expresses deep gratitude to Priya for her unwavering love and to his brother, Yash, for his constant support. Shubham also acknowledges the profound impact of his mentors, Avnish Aggarwal, Yogesh Dhandekar, and Amit Jain, whose guidance has been instrumental in shaping his professional journey. Their mentorship has not only honed his technical skills but also enriched his approach to complex problem-solving in the tech industry.

Table of Contents

3

Leveraging Behavioral Design Patterns   53

Part 2: Architecture and UI Patterns

4

Exploring Reactive View Library Patterns   83

5

Rendering Strategies and Page Hydration   111

6

Micro Frontends, Zones, and Islands Architectures   145

Part 3: Performance and Security Patterns

10

Asset Loading Strategies and Executing Code off the Main Thread   261

Preface

Welcome! JavaScript design patterns are techniques that allow us to write more robust, scalable, and extensible applications in JavaScript. JavaScript is the main programming language available in web browsers and is one of the most popular programming languages with support beyond browsers.

Design patterns are solutions to common problems that can be reused. The most-written-about design patterns come from the world of object-oriented programming.

JavaScript's attributes as a lightweight, multi-paradigm, dynamic, single-threaded language give it different strengths and weaknesses to other mainstream programming languages. It's common for software engineers to use JavaScript in addition to being well versed in a different programming language. JavaScript's different gearing means that implementing design patterns verbatim can lead to non-idiomatic and under-performing JavaScript applications.

There are many resources on JavaScript and design patterns, but this book provides a cohesive and comprehensive view of design patterns in modern (ECMAScript 6+) JavaScript with real-world examples of how to deploy them in a professional setting. In addition to this complete library of patterns to apply to projects, this book also provides an overview of how to structure different parts of an application to deliver high performance at scale.

In this book, you will be provided with up-to-date guidance through the world of modern JavaScript patterns based on nine years of experience building and deploying JavaScript and React applications at scale at companies such as Elsevier, Canon, and Eurostar, delivering multiple system evolutions, performance projects, and a next-generation frontend application architecture.

Who this book is for

This book is for developers and software architects who want to leverage JavaScript and the web platform to increase productivity, software quality, and the performance of their applications.

Familiarity with software design patterns would be a plus but is not required.

The three main challenges faced by developers and architects who are the target audience of this content are as follows:

- They are familiar with programming concepts but not how to effectively implement them in JavaScript
- They want to structure JavaScript code and applications in a way that is maintainable and extensible
- They want to deliver more performance to the users of their JavaScript applications

What this book covers

Chapter 1, Working with Creational Design Patterns, covers creational design patterns, which help to organize object creation. We'll look at implementing the prototype, singleton, and factory patterns in JavaScript.

Chapter 2, Implementing Structural Design Patterns, looks at structural design patterns, which help to organize relationships between entities. We'll implement the proxy, decorator, flyweight, and adapter patterns in JavaScript.

Chapter 3, Leveraging Behavioral Design Patterns, delves into behavioral design patterns, which help to organize communication between objects. We'll learn about the observer, state, strategy, and visitor patterns in JavaScript.

Chapter 4, Exploring Reactive View Library Patterns, explores reactive view libraries, such as React, which have taken over the JavaScript application landscape. With these libraries come new patterns to explore, implement, and contrast.

Chapter 5, Rendering Strategies and Page Hydration, takes a look at optimizing page performance, which is a key concern nowadays. It's a concern both for improving the on-page conversion of customers and search engine optimization, since search engines such as Google take core web vitals into account.

Chapter 6, Micro Frontends, Zones, and Islands Architectures, explores micro frontends. Akin to the microservices movement in the service tier, micro frontends are designed to split a large surface area into smaller chunks that can be worked on and delivered at higher velocity.

Chapter 7, Asynchronous Programming Performance Patterns, looks at how JavaScript's single-threaded event-loop-based concurrency model is one of its greatest strengths but is often misunderstood or under-leveraged in performance-sensitive situations. Writing asynchronous-handling code in JavaScript in a performant and extensible manner is key to delivering a smooth user experience at scale.

Chapter 8, Event-Driven Programming Patterns, explores how event-driven programming in JavaScript is of paramount importance in security-sensitive applications as it is a way to pass information from and to different web contexts. Event-driven applications can often be optimized to enable better performance and scalability.

Chapter 9, Maximizing Performance – Lazy Loading and Code Splitting, deals with how, in order to maximize the performance of a JavaScript application, reducing the amount of unused JavaScript being loaded and interpreted is key. The techniques that can be brought to bear on this problem are called lazy loading and code splitting.

Chapter 10, Asset-Loading Strategies and Executing Code off the Main Thread, looks at how there are situations in the lifecycle of an application where loading more JavaScript or assets is inevitable. You will learn about asset-loading optimizations in the specific case of JavaScript, as well as other web resources, and finally how to execute JavaScript off the main browser thread.

To get the most out of this book

You will need to have prior experience with JavaScript and developing for the web. Some of the more advanced topics in the book will be of interest to developers with intermediate experience in building for the web with JavaScript.

Software/hardware covered in the book	Operating system requirements
Node.js 20+	Windows, macOS, or Linux
NPM v8+	Windows, macOS, or Linux
ECMAScript 6+	Windows, macOS, or Linux
React v16+	Windows, macOS, or Linux
Next.js	Windows, macOS, or Linux

If you are using the digital version of this book, we advise you to type the code yourself or access the code from the book's GitHub repository (a link is available in the next section). Doing so will help you avoid any potential errors related to the copying and pasting of code.

Download the example code files

You can download the example code files for this book from GitHub at `https://github.com/PacktPublishing/JavaScript-Design-Patterns`. If there's an update to the code, it will be updated in the GitHub repository.

We also have other code bundles from our rich catalog of books and videos available at `https://github.com/PacktPublishing/`. Check them out!

Conventions used

There are a number of text conventions used throughout this book.

`Code in text`: Indicates code words in text, database table names, folder names, filenames, file extensions, pathnames, dummy URLs, user input, and Twitter handles. Here is an example: " In order to make the code easier to follow, we'll switch on the lowercased version of `tagName`."

A block of code is set as follows:

```
<script>
  // handle receiving messages from iframe -> parent
  const allowedMessageOrigins = ['http://127.0.0.1:8000'];
  window.addEventListener('message', (event) => {
    if (!allowedMessageOrigins.includes(event.origin)) {
      console.warn(
        `Dropping message due to non-allowlisted origin ${event.
origin}`,
        event,
      );
      return;
    }
    // no change to the rest of the message handler
  });
</script>
```

Bold: Indicates a new term, an important word, or words that you see onscreen. For instance, words in menus or dialog boxes appear in **bold**. Here is an example: "When opening the select, things seem to work ok, we're seeing the **Fruit:** prefix for all the options."

> **Tips or important notes**
> Appear like this.

Get in touch

Feedback from our readers is always welcome.

General feedback: If you have questions about any aspect of this book, email us at customercare@ packtpub.com and mention the book title in the subject of your message.

Errata: Although we have taken every care to ensure the accuracy of our content, mistakes do happen. If you have found a mistake in this book, we would be grateful if you would report this to us. Please visit www.packtpub.com/support/errata and fill in the form.

Piracy: If you come across any illegal copies of our works in any form on the internet, we would be grateful if you would provide us with the location address or website name. Please contact us at copyright@packt.com with a link to the material.

If you are interested in becoming an author: If there is a topic that you have expertise in and you are interested in either writing or contributing to a book, please visit authors.packtpub.com.

Share Your Thoughts

Once you've read, we'd love to hear your thoughts! Scan the QR code below to go straight to the Amazon review page for this book and share your feedback.

https://packt.link/r/1804612278

Your review is important to us and the tech community and will help us make sure we're delivering excellent quality content.

Download a free PDF copy of this book

Thanks for purchasing this book!

Do you like to read on the go but are unable to carry your print books everywhere?

Is your eBook purchase not compatible with the device of your choice?

Don't worry, now with every Packt book you get a DRM-free PDF version of that book at no cost.

Read anywhere, any place, on any device. Search, copy, and paste code from your favorite technical books directly into your application.

The perks don't stop there, you can get exclusive access to discounts, newsletters, and great free content in your inbox daily

Follow these simple steps to get the benefits:

1. Scan the QR code or visit the link below

https://packt.link/free-ebook/978-1-80461-227-9

2. Submit your proof of purchase
3. That's it! We'll send your free PDF and other benefits to your email directly

Part 1: Design Patterns

In this part, you will get an overview of design patterns and how they can be implemented effectively in modern JavaScript. You will learn how and when to implement creational, structural, and behavioral design patterns in the "classical" object-oriented way and how modern JavaScript features can be used to make this implementation more idiomatic to the language. Finally, you'll see real-world examples of design patterns being applied in the JavaScript ecosystem, thereby learning how to recognize them.

This part has the following chapters:

- *Chapter 1, Working with Creational Design Patterns*

- *Chapter 2, Implementing Structural Design Patterns*

- *Chapter 3, Leveraging Behavioral Design Patterns*

1
Working with Creational Design Patterns

JavaScript design patterns are techniques that allow us to write more robust, scalable, and extensible applications in JavaScript. JavaScript is a very popular programming language, in part due to its place as a way to deliver interactive functionality on web pages. The other reason for its popularity is JavaScript's lightweight, dynamic, multi-paradigm nature, which means that design patterns from other ecosystems can be adapted to take advantage of JavaScript's strengths. JavaScript's specific strengths and weaknesses can also inform new patterns specific to the language and the contexts in which it's used.

Creational design patterns give structure to object creation, which enables the development of systems and applications where different modules, classes, and objects don't need to know how to create instances of each other. The design patterns most relevant to JavaScript – the prototype, singleton, and factory patterns – will be explored, as well as situations where they're helpful and how to implement them in an idiomatic fashion.

We'll cover the following topics in this chapter:

- A comprehensive definition of creational design patterns and definitions for the prototype, singleton, and factory patterns
- Multiple implementations of the prototype pattern and its use cases
- An implementation of the singleton design pattern, eager and lazy initialization, use cases for singleton, and what a singleton pattern in modern JavaScript looks like
- How to implement the factory pattern using classes, a modern JavaScript alternative, and use cases

By the end of this chapter, you'll be able to identify when a creational design pattern is useful and make an informed decision on which of its multiple implementations to use, ranging from a more idiomatic JavaScript form to a classical form.

What are creational design patterns?

Creational design patterns handle object creation. They allow a consumer to create object instances without knowing the details of how to instantiate the object. Since, in object-oriented languages, instantiation of objects is limited to a class's constructor, allowing object instances to be created without calling the constructor is useful to reduce noise and tight coupling between the consumer and the class being instantiated.

In JavaScript, there's ambiguity when we discuss "object creation," since JavaScript's multi-paradigm nature means we can create objects without a class or a constructor. For example, in JavaScript this is an object creation using an object literal – `const config = { forceUpdate: true }`. In fact, modern idiomatic JavaScript tends to lean more toward procedural and function paradigms than object orientation. This means that creational design patterns may have to be adapted to be fully useful in JavaScript.

In summary, creational design patterns are useful in object-oriented JavaScript, since they hide instantiation details from consumers, which keeps coupling low, thereby allowing better module separation.

In the next section, we'll encounter our first creational design pattern – the prototype design pattern.

Implementing the prototype pattern in JavaScript

Let's start with a definition of the prototype pattern first.

The prototype design pattern allows us to create an instance based on another existing instance (our prototype).

In more formal terms, a `prototype` class exposes a `clone()` method. Consuming code, instead of calling `new SomeClass`, will call `new SomeClassPrototype(someClassInstance).clone()`. This method call will return a `new SomeClass` instance with all the values copied from `someClassInstance`.

Implementation

Let's imagine a scenario where we're building a chessboard. There are two key types of squares – white and black. In addition to this information, each square contains information such as its row, file, and which piece sits atop it.

A `BoardSquare` class constructor might look like the following:

```
class BoardSquare {
  constructor(color, row, file, startingPiece) {
    this.color = color;
    this.row = row;
```

```
      this.file = file;
    }
  }
```

A set of useful methods on BoardSquare might be occupySquare and clearSquare, as follows:

```
class BoardSquare {
  // no change to the rest of the class
  occupySquare(piece) {
    this.piece = piece;
  }
  clearSquare() {
    this.piece = null;
  }
}
```

Instantiating BoardSquare is quite cumbersome, due to all its properties:

```
const whiteSquare = new BoardSquare('white');
const whiteSquareTwo = new BoardSquare('white');
// ...
const whiteSquareLast = new BoardSquare('white');
```

Note the repetition of arguments being passed to new BoardSquare, which will cause issues if we want to change all board squares to black. We would need to change the parameter passed to each call of BoardSquare is one by one for each new BoardSquare call. This can be quite error-prone; all it takes is one hard-to-find mistake in the color value to cause a bug:

```
const blackSquare = new BoardSquare('black');
const blackSquareTwo = new BoardSquare('black');
// ...
const blackSquareLast = new BoardSquare('black');
```

Implementing our instantiation logic using a classical prototype looks as follows. We need a BoardSquarePrototype class; its constructor takes a prototype property, which it stores on the instance. BoardSquarePrototype exposes a clone() method that takes no arguments and returns a BoardSquare instance, with all the properties of prototype copied onto it:

```
class BoardSquarePrototype {
  constructor(prototype) {
    this.prototype = prototype;
  }
  clone() {
    const boardSquare = new BoardSquare();
    boardSquare.color = this.prototype.color;
```

```
    boardSquare.row = this.prototype.row;
    boardSquare.file = this.prototype.file;
    return boardSquare;
  }
}
```

Using `BoardSquarePrototype` requires the following steps:

1. First, we want an instance of `BoardSquare` to initialize – in this case, with `'white'`. It will then be passed as the `prototype` property during the `BoardSquarePrototype` constructor call:

    ```
    const whiteSquare = new BoardSquare('white');
    const whiteSquarePrototype = new BoardSquarePrototype
      (whiteSquare);
    ```

2. We can then use `whiteSquarePrototype` with `.clone()` to create our copies of `whiteSquare`. Note that `color` is copied over but each call to `clone()` returns a new instance.

    ```
    const whiteSquareTwo = whiteSquarePrototype.clone();
    // ...
    const whiteSquareLast = whiteSquarePrototype.clone();

    console.assert(
      whiteSquare.color === whiteSquareTwo.color &&
        whiteSquareTwo.color === whiteSquareLast.color,
      'Prototype.clone()-ed instances have the same color
        as the prototype'
    );
    console.assert(
      whiteSquare !== whiteSquareTwo &&
        whiteSquare !== whiteSquareLast &&
        whiteSquareTwo !== whiteSquareLast,
      'each Prototype.clone() call outputs a different
        instances'
    );
    ```

Per the assertions in the code, the cloned instances contain the same value for `color` but are different instances of the `Square` object.

A use case

To illustrate what it would take to change from a white square to a black square, let's look at some sample code where 'white' is not referenced in the variable names:

```
const boardSquare = new BoardSquare('white');
const boardSquarePrototype = new BoardSquarePrototype(boardSquare);

const boardSquareTwo = boardSquarePrototype.clone();
// ...
const boardSquareLast = boardSquarePrototype.clone();

console.assert(
  boardSquareTwo.color === 'white' &&
    boardSquare.color === boardSquareTwo.color &&
    boardSquareTwo.color === boardSquareLast.color,
  'Prototype.clone()-ed instances have the same color as
    the prototype'
);
console.assert(
  boardSquare !== boardSquareTwo &&
    boardSquare !== boardSquareLast &&
    boardSquareTwo !== boardSquareLast,
  'each Prototype.clone() call outputs a different
    instances'
);
```

In this scenario, we would only have to change the color value passed to BoardSquare to change the color of all the instances cloned from the prototype:

```
const boardSquare = new BoardSquare('black');
// rest of the code stays the same
console.assert(
  boardSquareTwo.color === 'black' &&
    boardSquare.color === boardSquareTwo.color &&
    boardSquareTwo.color === boardSquareLast.color,
  'Prototype.clone()-ed instances have the same color as
    the prototype'
);
console.assert(
  boardSquare !== boardSquareTwo &&
    boardSquare !== boardSquareLast &&
    boardSquareTwo !== boardSquareLast,
  'each Prototype.clone() call outputs a different
```

```
        instances'
    );
```

The prototype pattern is useful in situations where a "template" for the object instances is useful. It's a good pattern to create a "default object" but with custom values. It allows faster and easier changes, since they are implemented once on the template object but are applied to all clone() -ed instances.

Increasing robustness to change in the prototype's instance variables with modern JavaScript

There are improvements we can make to our prototype implementation in JavaScript.

The first is in the clone() method. To make our prototype class robust to changes in the prototype's constructor/instance variables, we should avoid copying the properties one by one.

For example, if we add a new startingPiece parameter that the BoardSquare constructor takes and sets the piece instance variable to, our current implementation of BoardSquarePrototype will fail to copy it, since it only copies color, row, and file:

```
class BoardSquare {
  constructor(color, row, file, startingPiece) {
    this.color = color;
    this.row = row;
    this.file = file;
    this.piece = startingPiece;
  }
  // same rest of the class
}

const boardSquare = new BoardSquare('white', 1, 'A',
  'king');
const boardSquarePrototype = new BoardSquarePrototype
  (boardSquare);
const otherBoardSquare = boardSquarePrototype.clone();

console.assert(
  otherBoardSquare.piece === undefined,
  'prototype.piece was not copied over'
);
```

> **Note**
> Reference for Object.assign: https://developer.mozilla.org/en-US/docs/Web/JavaScript/Reference/Global_Objects/Object/assign.

If we amend our `BoardSquarePrototype` class to use `Object.assign(new BoardSquare(), this.prototype)`, it will copy all the enumerable properties of `prototype`:

```
class BoardSquarePrototype {
  constructor(prototype) {
    this.prototype = prototype;
  }
  clone() {
    return Object.assign(new BoardSquare(), this.prototype);
  }
}

const boardSquare = new BoardSquare('white', 1, 'A',
  'king');
const boardSquarePrototype = new BoardSquarePrototype
  (boardSquare);
const otherBoardSquare = boardSquarePrototype.clone();

console.assert(
  otherBoardSquare.piece === 'king' &&
    otherBoardSquare.piece === boardSquare.piece,
  'prototype.piece was copied over'
);
```

The prototype pattern without classes in JavaScript

For historical reasons, JavaScript has a prototype concept deeply embedded into the language. In fact, classes were introduced much later into the ECMAScript standard, with ECMAScript 6, which was released in 2015 (for reference, ECMAScript 1 was published in 1997).

This is why a lot of JavaScript completely forgoes the use of classes. The JavaScript "object prototype" can be used to make objects inherit methods and variables from each other.

One way to clone objects is by using the `Object.create` to clone objects with their methods. This relies on the JavaScript prototype system:

```
const square = {
  color: 'white',
  occupySquare(piece) {
    this.piece = piece;
  },
  clearSquare() {
    this.piece = null;
  },
```

```
};
const otherSquare = Object.create(square);
```

One subtlety here is that `Object.create` does not actually copy anything; it simply creates a new object and sets its prototype to `square`. This means that if properties are not found on `otherSquare`, they're accessed on `square`:

```
console.assert(otherSquare.__proto__ === square, 'uses JS
  prototype');

console.assert(
  otherSquare.occupySquare === square.occupySquare &&
    otherSquare.clearSquare === square.clearSquare,
  "methods are not copied, they're 'inherited' using the
    prototype"
);

delete otherSquare.color;
console.assert(
  otherSquare.color === 'white' && otherSquare.color ===
    square.color,
  'data fields are also inherited'
);
```

A further note on the JavaScript prototype, and its existence before classes were part of JavaScript, is that subclassing in JavaScript is another syntax for setting an object's prototype. Have a look at the following `extends` example. `BlackSquare extends Square` sets the `prototype.__proto__` property of `BlackSquare` to `Square.prototype`:

```
class Square {
  constructor() {}
  occupySquare(piece) {
    this.piece = piece;
  }
  clearSquare() {
    this.piece = null;
  }
}

class BlackSquare extends Square {
  constructor() {
    super();
    this.color = 'black';
  }
```

```
  }

console.assert(
  BlackSquare.prototype.__proto__ === Square.prototype,
  'subclass prototype has prototype of superclass'
);
```

In this section, we learned how to implement the prototype pattern with a prototype class that exposes a `clone()` method, which code situations the prototype patterns can help with, and how to further improve our prototype implementation with modern JavaScript features. We also covered the JavaScript "prototype," why it exists, and its relationship with the prototype design pattern.

In the next part of the chapter, we'll look at another creational design pattern, the singleton design pattern, with some implementation approaches in JavaScript and its use cases.

The singleton pattern with eager and lazy initialization in JavaScript

To begin, let's define the singleton design pattern.

The singleton pattern allows an object to be instantiated only once, exposes this single instance to consumers, and controls the instantiation of the single instance.

The singleton is another way of getting access to an object instance without using a constructor, although it's necessary for the object to be designed as a singleton.

Implementation

A classic example of a singleton is a logger. It's rarely necessary (and often, it's a problem) to instantiate multiple loggers in an application. Having a singleton means the initialization site is controlled, and the logger configuration will be consistent across the application – for example, the log level won't change depending on where in the application we call the logger from.

A simple logger looks something as follows, with a constructor taking `logLevel` and `transport`, and an `isLevelEnabled` private method, which allows us to drop logs that the logger is not configured to keep (for example, when the level is `warn` we drop `info` messages). The logger finally implements the `info`, `warn`, and `error` methods, which behave as previously described; they only call the relevant `transport` method if the level is "enabled" (i.e., "above" what the configured log level is).

The possible `logLevel` values that power `isLevelEnabled` are stored as a static field on `Logger`:

```
class Logger {
  static logLevels = ['info', 'warn', 'error'];
  constructor(logLevel = 'info', transport = console) {
```

```
    if (Logger.#loggerInstance) {
      throw new TypeError(
        'Logger is not constructable, use getInstance()
          instead'
      );
    }
    this.logLevel = logLevel;
    this.transport = transport;
  }
  isLevelEnabled(targetLevel) {
    return (
      Logger.logLevels.indexOf(targetLevel) >=
      Logger.logLevels.indexOf(this.logLevel)
    );
  }
  info(message) {
    if (this.isLevelEnabled('info')) {
      return this.transport.info(message);
    }
  }
  warn(message) {
    if (this.isLevelEnabled('warn')) {
      this.transport.warn(message);
    }
  }
  error(message) {
    if (this.isLevelEnabled('error')) {
      this.transport.error(message);
    }
  }
}
```

In order to make Logger a singleton, we need to implement a getInstance static method that returns a cached instance. In order to do, this we'll use a static loggerInstance on Logger. getInstance will check whether Logger.loggerInstance exists and return it if it does; otherwise, it will create a new Logger instance, set that as loggerInstance, and return it:

```
class Logger {
  static loggerInstance = null;
  // rest of the class
  static getInstance() {
    if (!Logger.loggerInstance) {
      Logger.loggerInstance = new Logger('warn', console);
    }
```

```
    return Logger.loggerInstance;
  }
}
```

Using this in another module is as simple as calling `Logger.getInstance()`. All `getInstance` calls will return the same instance of `Logger`:

```
const a = Logger.getInstance();
const b = Logger.getInstance();
console.assert(a === b, 'Logger.getInstance() returns the
  same reference');
```

We've implemented a singleton with "lazy" initialization. The initialization occurs when the first `getInstance` call is made. In the next section, we'll see how we might extend our code to have an "eager" initialization of `loggerInstance`, where `loggerInstance` will be initialized when the `Logger` code is evaluated.

Ensuring only one singleton instance is constructed

A characteristic of a singleton is the "single instance" concept. We want to "force" consumers to use the `getInstance` method.

In order to do this, we can check for the existence of `loggerInstance` when the contructor is called:

```
class Logger {
  // rest of the class
  constructor(logLevel = 'info', transport = console) {
    if (Logger.loggerInstance) {
      throw new TypeError(
        'Logger is not constructable, use getInstance()
          instead'
      );
    }
    this.logLevel = logLevel;
    this.transport = transport;
  }
  // rest of the class
}
```

In the case where we call `getInstance` (and, therefore, `Logger.loggerInstance` is populated), the constructor will now throw an error:

```
Logger.getInstance();
new Logger('info', console); // new TypeError('Logger is
  not constructable, use getInstance() instead');
```

This behavior is useful to ensure that consumers don't instantiate their own Logger and they use getInstance instead. All consumers using getInstance means the configuration to set up the logger is encapsulated by the Logger class.

There's still a gap in the implementation, as constructing new Logger() before any getInstance() calls will succeed, as shown in the following example:

```
new Logger('info', console); // Logger { logLevel: 'info',
  transport: ... }
new Logger('info', console); // Logger { logLevel: 'info',
  transport: ... }
Logger.getInstance();
new Logger('info', console); // new TypeError('Logger is
  not constructable, use getInstance() instead');
```

In multithreaded languages, our implementation would also have a potential race condition – multiple consumers calling Logger.getInstance() concurrently could cause multiple instances to exist. However, since popular JavaScript runtimes are single-threaded, we won't have to worry about such a race condition – getInstance is a "synchronous" method, so multiple calls to it would be interpreted one after the other. For reference, Node.js, Deno, and the mainstream browsers Chrome, Safari, Edge, and Firefox provide a single-threaded JavaScript runtime.

Singleton with eager initialization

Eager initialization can be useful to ensure that the singleton is ready for use and features, such as disabling the constructor when an instance exists, work for all cases.

We can eager-initialize by setting Logger.loggerInstance in the Logger constructor:

```
class Logger {
  // rest of the class unchanged
  constructor(logLevel = 'info', transport = console) {
    // rest of the constructor unchanged
    Logger.loggerInstance = this;
  }
}
```

This approach has the downside of the constructor performing a global state mutation, which isn't ideal from a "single responsibility principle" standpoint; the constructor now has a side-effect of sorts (mutating global state) beyond its responsibility to set up an object instance.

An alternative way to eager-initialize is by running `Logger.getInstance()` in the logger's module; it's useful to pair it with an `export default` statement:

```
export class Logger {
  // no changes to the Logger class
}
export default Logger.getInstance();
```

With the preceding exports added, there are now two ways to access a logger instance. The first is to import `Logger` by name and call `Logger.getInstance()`:

```
import { Logger } from './logger.js';
const logger = Logger.getInstance();
logger.warn('testing testing 12'); // testing testing 12
```

The second way to use the logger is by importing the default export:

```
import logger from './logger.js';
logger.warn('testing testing 12'); // testing testing 12
```

Any code now importing `Logger` will get a pre-determined singleton instance of the logger.

Use cases

A singleton shines when there should only be one instance of an object in an application – for example, a logger that shouldn't be set up/torn down on every request.

Since the singleton class controls how it gets instantiated, it's also a good fit for objects that are tricky to configure (again, a logger, a metrics exporter, and an API client are good examples). The instantiation is completely encapsulated if, like in our example, we "disable" the constructor.

There's a performance benefit to constraining the application to a single instance of an object in terms of memory footprint.

The major drawbacks of singletons are an effect of their reliance on global state (in our example, the static `loggerInstance`). It's hard to test a singleton, especially in a case where the constructor is "disabled" (like in our example), since our tests will want to always have a single instance of the singleton.

Singletons can also be considered "global state" to some extent, which comes with all its drawbacks. Global state can sometimes be a sign of poor design, and updating/consuming global state is error-prone (e.g., if a consumer is reading state but it is then updated and not read again).

Improvements with the "class singleton" pattern

With our singleton logger implementation, it's possible to modify the internal state of the singleton from outside of it. This is nothing specific to our singleton; it's the nature of JavaScript. By default, its fields and methods are public.

However, this is a bigger issue in our singleton scenario, since a consumer could reset `loggerInstance` using a statement such as `Logger.loggerInstance = null` or `delete Logger.loggerInstance`. See the following example:

```
const logger = Logger.getInstance();
Logger.loggerInstance = null;
const logger = new Logger('info', console); // should throw but
creates a new instance
```

In order to stop consumers from modifying the `loggerInstance` static field, we can make it a private field. Private fields in JavaScript are part of the ECMAScript 2023 specification (the 13th ECMAScript edition).

To define a private field, we use the # prefix for the field name – in this case, `loggerInstance` becomes `#loggerInstance`. The `isLevelEnabled` method becomes `#isLevelEnabled`, and we also declare `logLevel` and `transport` as `#logLevel` and `#transport`, respectively:

```
export class Logger {
  // other static fields are unchanged
  static #loggerInstance = null;
  #logLevel;
  #transport;
  constructor(logLevel = 'info', transport = console) {
    if (Logger.#loggerInstance) {
      throw new TypeError(
        'Logger is not constructable, use getInstance()
          instead'
      );
    }
    this.#logLevel = logLevel;
    this.#transport = transport;
  }
  #isLevelEnabled(targetLevel) {
    // implementation unchanged
  }
  info(message) {
    if (this.#isLevelEnabled('info')) {
      return this.#transport.info(message);
    }
```

```
  }
  warn(message) {
    if (this.#isLevelEnabled('warn')) {
      this.#transport.warn(message);
    }
  }
  error(message) {
    if (this.#isLevelEnabled('error')) {
      this.#transport.error(message);
    }
  }
  getInstance() {
    if (!Logger.#loggerInstance) {
      Logger.#loggerInstance = new Logger('warn', console);
    }

    return Logger.#loggerInstance;
  }
}
```

It's not possible to delete `loggerInstance` or set it to `null`, since attempting to access `Logger.#loggerInstance` is a syntax error:

```
Logger.#loggerInstance = null;
       ^
```

```
SyntaxError: Private field '#loggerInstance' must be
  declared in an enclosing class
```

Another useful technique is to disallow modification of fields on an object. In order to disallow modification, we can use `Object.freeze` to freeze the instance once it's created.

```
class Logger {
  // no changes to the logger class
}
export default Object.freeze(new Logger('warn', console));
```

Now, when someone attempts to change a field on the `Logger` instance, they'll get `TypeError`:

```
import logger from './logger.js';
logger.transport = {}; // new TypeError('Cannot add
  property transport, object is not extensible')
```

We've now refactored our singleton implementation to disallow external modifications to it by using private fields and `Object.freeze`. Next, we'll see how to use **EcmaScript (ES)** modules to deliver singleton functionality.

A singleton without class fields using ES module behavior

The JavaScript module system has the following caching behavior – if a module is loaded, any further imports of the module's exports will be cached instances of exports.

Therefore, it's possible to create a singleton as follows in JavaScript.

```
class MySingleton {
  constructor(value) {
    this.value = value;
  }
}
export default new MySingleton('my-value');
```

Multiple imports of the default export will result in only one existing instance of the `MySingleton` object. Furthermore, if we don't export the class, then the constructor doesn't need to be "protected."

As the following snippet with dynamic imports shows, both `import('./my-singleton.js')` result in the same object. They both return the same object because the output of the `import` for a given module is a singleton:

```
await Promise.all([
  import('./my-singleton.js'),
  import('./my-singleton.js'),
]).then(([import1, import2]) => {
  console.assert(
    import1.default.value === 'my-value' &&
      import2.default.value === 'my-value',
    'instance variable is equal'
  );
  console.assert(
    import1.default === import2.default,
    'multiple imports of a module yield the same default
      object value, a single MySingleton instance'
  );
  console.assert(import1 === import2, 'import objects are a
    single reference');
});
```

For our logger, this means we could implement an eager-initialized singleton in JavaScript without any of the heavy-handed guarding of the constructor or even a `getInstance` method. Note the use of `logLevel` and `isLevelEnabled` as a public instance property and a public method, respectively (since it might be useful to have access to them from a consumer). In the meantime, `#transport` remains private, and we've dropped `loggerInstance` and `getInstance`. We've kept `Object.freeze()`, which means that even though `logLevel` is readable from a consumer, it's not available to modify:

```
class Logger {
  static logLevels = ['info', 'warn', 'error'];
  #transport;
  constructor(logLevel = 'info', transport = console) {
    this.logLevel = logLevel;
    this.#transport = transport;
  }
  isLevelEnabled(targetLevel) {
    return (
      Logger.logLevels.indexOf(targetLevel) >=
      Logger.logLevels.indexOf(this.logLevel)
    );
  }
  info(message) {
    if (this.isLevelEnabled('info')) {
      return this.#transport.info(message);
    }
  }
  warn(message) {
    if (this.isLevelEnabled('warn')) {
      this.#transport.warn(message);
    }
  }
  error(message) {
    if (this.isLevelEnabled('error')) {
      this.#transport.error(message);
    }
  }
}

export default Object.freeze(new Logger('warn', console));
```

In this part of the chapter, we learned how to implement the singleton pattern with a class that exposes a `getInstance()` method, as well as the difference between the eager and lazy initialization of a singleton. We've covered some JavaScript features, such as private class fields and `Object.freeze`, which can be useful when implementing the singleton pattern. Finally, we explored how JavaScript/ECMAScript modules have singleton-like behavior and can be relied upon to provide this behavior for a class instance.

In the next section, we'll explore the final creational design pattern covered in this chapter – the factory design pattern.

The factory pattern in JavaScript

In a similar fashion to the discussion about the JavaScript "prototype" versus the prototype creational design pattern, "factory" refers to related but different concepts when it comes to general program design discussions and design patterns.

A "factory," in the general programming sense, is an object that's built with the goal of creating other objects. This is hinted at by the name that refers to a facility that processes items from one shape into another (or from one type of item to another). This factory denomination means that the output of a function or method is a new object. In JavaScript, this means that something as simple as a function that returns an object literal is a factory function:

```
const simpleFactoryFunction = () => ({}); // returns an object,
therefore it's a factory.
```

This definition of a factory is useful, but this section of the chapter is about the factory design pattern, which does fit into this overall "factory" definition.

The factory or factory method design pattern solves a class inheritance problem. A base or superclass is extended (the extended class is a subclass). The base class's role is to provide orchestration for the methods implemented in the subclasses, as we want the subclasses to control which other objects to populate an instance with.

Implementation

A factory example is as follows. We have a `Building` base class that implements a `generateBuilding()` method. For now, it's going to create a top floor using the `makeTopFloor` instance method. In the base class (`Building`), `makeTopFloor` is implemented, mainly because JavaScript doesn't provide a way to define abstract methods. The `makeTopFloor` implementation throws an error because subclasses should override it; `makeTopFloor` is the "factory method" in this case. It's how the base class defers the instantiation of objects to the subclasses:

```
class Building {
  generateBuilding() {
```

```
    this.topFloor = this.makeTopFloor();
  }
  makeTopFloor() {
    throw new Error('not implemented, left for subclasses
      to implement');
  }
}
```

If we wanted to implement a single-story house, we would extend `Building` and override `makeTopFloor`; in this instance, `topFloor` will have `level: 1`.

```
class House extends Building {
  makeTopFloor() {
    return {
      level: 1,
    };
  }
}
```

When we instantiate `House`, which is a subclass of `Building`, we have access to the `generateBuilding` method; when called, it sets `topFloor` correctly (to `{ level: 1 }`).

```
const house = new House();
house.generateBuilding();
console.assert(house.topFloor.level === 1, 'topFloor works
  in House');
```

Now, if we want to create a different type of building that has a very different top floor, we can still extend `Building`; we simply override `makeTopFloor` to return a different floor. In the case of a skyscraper, we want the top floor to be very high, so we'll do the following:

```
class SkyScraper extends Building {
  makeTopFloor() {
    return {
      level: 125,
    };
  }
}
```

Having defined our `SkyScraper`, which is a subclass of `Building`, we can instantiate it and call `generateBuilding`. As in the preceding `House` case, the `generateBuilding` method will use `SkyScraper`'s `makeTopFloor` method to populate the `topFloor` instance property:

```
const skyScraper = new SkyScraper();
skyScraper.generateBuilding();
console.assert(skyScraper.topFloor.level > 100, 'topFloor
  works in SkyScraper');
```

The "factory method" in this case is `makeTopFloor`. The `makeTopFloor` method is "not implemented" in the base class, in the sense that it's implemented in a manner that forces subclasses that wish to use `generateBuilding` to define a `makeTopFloor` override.

Note that `makeTopFloor` in our examples returned object literals, as mentioned earlier in the chapter; this is a feature of JavaScript not available in all object-oriented languages (JavaScript is multi-paradigm). We'll see different ways to implement the factory pattern later in this section.

Use cases

The benefit of using a factory method is that we can create a wide variety of subclasses without modifying the base class. This is the "open/closed principle" at play – the `Building` class in our example is "open" to extension (i.e., can be subclassed to infinity for different types of buildings) but "closed" to modification (i.e., we don't need to make changes in `Building` for every subclass, only when we want to add new behaviors).

Improvements with modern JavaScript

The key improvement we can make with JavaScript is enabled by its first-class support for functions and the ability to define objects using literals (instead of classes being instantiated).

JavaScript having "first-class functions" means functions are like any other type – they can be passed as parameters, set as variable values, and returned from other functions.

A more idiomatic implementation of this pattern would probably involve a `generateBuilding` standalone function instead of a `Building` class. `generateBuilding` would take `makeTopFloor` either as a parameter or take an object parameter with a `makeTopFloor` key. The output of `generateBuilding` would be an object created using an object literal, which takes the output of `makeTopFloor()` and sets it as the value to a `topFloor` key:

```
function generateBuilding({ makeTopFloor }) {
  return {
    topFloor: makeTopFloor(),
  };
}
```

In order to create our house and skyscraper, we would call `generateBuilding` with the relevant `makeTopFloor` functions. In the case of the house, we want a top floor that is on level 1; in the case of the skyscraper, we want a top floor on level 125.

```
const house = generateBuilding({
  makeTopFloor() {
    return {
      level: 1,
    };
  },
});
console.assert(house.topFloor.level === 1, 'topFloor works
  in house');

const skyScraper = generateBuilding({
  makeTopFloor() {
    return {
      level: 125,
    };
  },
});
console.assert(skyScraper.topFloor.level > 100, 'topFloor works in
skyScraper');
```

One reason why using functions directly works better in JavaScript is that we didn't have to implement a "throw an error to remind consumers to override me" `makeFloor` method that we had with the `Building` class.

In languages other than JavaScript that have support for abstract methods, this pattern is more useful and natural to implement than in JavaScript, where we have first-class functions.

You also have to bear in mind that the original versions of JavaScript/ECMAScript didn't include a `class` construct.

In the final section of the chapter, we learned what the factory method pattern is and how it contrasts with the factory programming concept. We then implemented a class-based factory pattern scenario as well as a more idiomatic JavaScript version. Interspersed through this section, we covered the use cases, benefits, and drawbacks of the factory method pattern in JavaScript.

Summary

Throughout this chapter, we discussed how creational design patterns allow us to build more extensible and maintainable systems in JavaScript.

The prototype design pattern shines when creating many instances of objects that contain the same values. This design pattern allows us to change the initial values of the prototype and affect all the cloned instances.

The singleton design pattern is useful to completely hide initialization details of a class that should really only be instantiated once. We saw how JavaScript's module system generates singletons and how that can be leveraged to simplify a singleton implementation.

The factory method design pattern allows a base class to defer the implementation of some object creations to subclasses. We saw which features would make this pattern more useful in JavaScript, as well as an alternative idiomatic JavaScript approach with factory functions.

We can now leverage creational design patterns to build classes that are composable and can be evolved as necessary to cover different use cases.

Now that we know how to create objects efficiently with creational design patterns, in the next chapter, we'll cover how to use structural design patterns to organize relationships between different objects and classes.

2
Implementing Structural Design Patterns

Structural design patterns give us tools to handle *connecting* different objects; in other words, managing the relationships between objects. This includes techniques to reduce memory usage and develop functionality with existing classes without modifying these existing classes. In addition, JavaScript features allow us to more effectively apply these patterns. Modern JavaScript includes some built-ins that allow us to implement structural design patterns in a more efficient manner.

We'll cover the following topics in this chapter:

- Defining structural design patterns as a whole, and proxy, decorator, flyweight, and adapter specifically

- An implementation of the proxy pattern with a class-based approach as well as an alternative using Proxy and Reflect

- Multiple implementations of the decorator pattern, leveraging JavaScript first-class support for functions

- An iterative approach to implementing flyweight in JavaScript, including ergonomic improvements using modern JavaScript features

- Class- and function-based adapter implementations

At the end of this chapter, you'll be able to make informed decisions on when and how to use structural design patterns in JavaScript.

Technical requirements

You can find the code files for this chapter on GitHub at `https://github.com/PacktPublishing/Javascript-Design-Patterns`

What are structural design patterns?

When building software, we want to be able to *connect* different pieces of code (e.g., classes and functions) and change how the parties involved in these connections and relationships interact without having to jump through multiple fragmented parts of the codebase.

Structural design patterns allow us to add, remove, and change functionality in modules and classes safely. The "structural" aspect of these patterns is due to the fact that we can play around with implementations if the exposed interfaces are stable.

Structural design patterns are a good way to maintain the separation of concerns and loose coupling of different classes and modules while maintaining a high development velocity.

In the next section, we'll look at multiple approaches to implement the Proxy pattern in JavaScript.

Implementing the Proxy pattern with Proxy and Reflect

The proxy pattern involves providing an object (the `subject`, or `real` object) that fulfills a certain interface. The `proxy` (a `placeholder` or `wrapper` object) controls access to the `subject`. This allows us to provide additional functionality on top of the subject without changing a consumer's interactions with the `subject`.

This means that a proxy needs to provide an interface matching the `subject`.

By using the proxy pattern, we can intercept all operations on the original object and either pass them through or change their implementation. This follows the open/closed principle, where both the `subject` and `consumer` are closed for modification, but the proxy provides us with a hook to `extend`, which means the design is open to extension.

A redaction proxy implementation

We'll start with the following implementation class that has a couple of methods that output strings:

```
class Implementation {
  someFn() {
    return 'some-output';
  }
  sensitiveFn() {
    return 'sensitive-output';
  }
}
```

Let's imagine that the `sensitive` string in the output should be redacted.

Here's how a `RedactionProxy` class could look:

```
class RedactionProxy {
  constructor() {
    this.impl = new Implementation();
  }
  someFn() {
    return this.impl.someFn();
  }
  sensitiveFn() {
    return this.impl.sensitiveFn().replace('sensitive',
      '[REDACTED]');
  }
}
```

In this case, `RedactionProxy` does what we call a **pass-through** of `someFn()` calls. In other words, `RedactionProxy#someFn` simply forwards the `someFn` call to `Implementation`. See the following illustration:

```
const redactionProxy = new RedactionProxy();
console.assert(
  redactionProxy.someFn() === newImplementation().someFn(),
    'Proxy implementation calls through to original'
);
```

When it comes to `sensitiveFn`, `RedactionProxy` implements the same interface as `Implementation` except it overrides the output, replacing `sensitive` with `[REDACTED]`.

This means the interface for `RedactionProxy` and `Implementation` is the same, but `RedactionProxy` can control which method calls and fields are available along with their implementation. See the following example of this behavior:

```
console.assert(
  redactionProxy.sensitiveFn() !== new
    Implementation().sensitiveFn()&&
      redactionProxy.sensitiveFn() === '[REDACTED]-output',
      'Proxy implementation adds new behaviour'
);
```

Use cases

The proxy pattern allows us to intercept calls to an object (the `implementation` or `subject`) and augment them, either by manipulating the output or by adding a side-effect.

Our example of redaction is a good use case for it, but any other type of instrumentation is also a good use case. The instrumentation could be concerned with measuring something about a function/field access (e.g. the time it takes) or ensuring access to a property triggers a certain effect. For example, the **reactivity** system of Vue.js and Alpine.js is based on proxies, where a JavaScript Proxy object is used to wrap the reactive data objects. This allows the library (Vue or Alpine) to detect when properties are changed and run things such as watchers, effects, and re-renders.

Improving the proxy pattern in JavaScript with the Proxy and Reflect global objects

Back to our example, what happens when we need to redact more functions?

Let's take an `Implementation` class with three methods (`someFn`, `sensitiveFn`, and `otherSensitiveFn`):

```
class Implementation {
  someFn() {
    return 'sensitive-some-output';
  }
  sensitiveFn() {
    return 'sensitive-output';
  }
  otherSensitiveFn() {
    return 'sensitive-other-output';
  }
}
```

A naïve implementation of an extended proxy looks as follows, where each method calls the implementation's method and then replaces `sensitive` in its output:

```
class RedactionProxyNaive {
  constructor() {
    this.impl = new Implementation();
  }
  someFn() {
    return this.impl.someFn().replace
      ('sensitive', '[REDACTED]');
  }
  sensitiveFn() {
    return this.impl.sensitiveFn().replace('sensitive',
      '[REDACTED]');
  }
  otherSensitiveFn() {
    return this.impl.otherSensitiveFn().
```

```
      replace('sensitive', '[REDACTED]');
  }
}
```

This implementation of the `Proxy` works, as we can ensure with the following code:

```
console.assert(
  !new RedactionProxyNaive().someFn().includes('sensitive')
    &&
    !new RedactionProxyNaive().sensitiveFn().includes
      ('sensitive') &&
    !new RedactionProxyNaive().otherSensitiveFn().includes
      ('sensitive'),
  'naive proxy redacts correctly'
);
```

One improvement we can make here is to extract a `#redact` private method to handle the replacement of `sensitive`:

```
class RedactionProxyNaiveRefactored {
  constructor() {
    this.impl = new Implementation();
  }
  #redact(str) {
    return str.replace('sensitive', '[REDACTED]');
  }
  someFn() {
    return this.#redact(this.impl.someFn());
  }
  sensitiveFn() {
    return this.#redact(this.impl.sensitiveFn());
  }
  otherSensitiveFn() {
    return this.#redact(this.impl.otherSensitiveFn());
  }
}

console.assert(
  !new RedactionProxyNaiveRefactored().someFn().includes
    ('sensitive') &&
    !new RedactionProxyNaiveRefactored().sensitiveFn().
      includes('sensitive') &&
    !new RedactionProxyNaiveRefactored()
      .otherSensitiveFn()
```

```
        .includes('sensitive'),
    'refactored naive proxy redacts correctly'
  );
```

The downside of this approach is that every method on the `Implementation` object (the subject) will require a change to our Proxy implementation.

Fortunately, JavaScript has a built-in class to programmatically manage these situations. The JavaScript class is aptly called `Proxy`.

Let's take the following plain JavaScript object (this also works for class instances) with both fields and functions:

```
const obj = {
  someFn() {
    return 'sensitive-some-output';
  },
  sensitiveFn() {
    return 'sensitive-output';
  },
  otherSensitiveFn() {
    return 'sensitive-other-output';
  },
  field: 'sensitive-data',
  sensitiveField: 'redact-everything',
};
```

We want to be able to completely redact (i.e., keep none of the original output) those fields that contain `sensitive` in the field or method name. We also want to have a value redaction functionality when the output contains the string `sensitive`, where we replace `sensitive` with `[REDACTED]`.

To achieve this, we define a Proxy that will wrap our `obj` object. We instantiate the Proxy with a "get trap," which allows us to intercept all property accesses (which includes function access).

The `get` function receives a `target` and `property`. The target is the object being wrapped (`obj`), `property` is the property being accessed.

Based on whether `target[property]` is a function or not, we'll replace it with a wrapper function that will collect all the arguments, call `target[property]` with those arguments, intercept the output, and replace `sensitive` with `[REDACTED]`. We also return `[REDACTED]` if the property name includes `sensitive` (in our case, using `sensitiveFn`).

In cases where `target[property]` is not a function, we'll do a full redaction if the property name includes `sensitive` and also replace `sensitive` in the output for all other properties:

```
const redactedObjProxy = new Proxy(obj, {
  get(target, property, _receiver) {
    if (target[property] instanceof Function) {
      return (...args) => {
        if (property.includes('sensitive')) {
          return '[REDACTED]';
        }
        const output = target[property](...args);
        if (typeof output === 'string') {
          return output.replace('sensitive', '[REDACTED]');
        }
        return output;
      };
    }
    if (property.includes('sensitive')) {
      return '[REDACTED]';
    }
    return target[property].replace('sensitive',
      '[REDACTED]');
  },
});
```

The following code ensures our Proxy implementation works as expected. `sensitive` is not present in any of the function output or in the `field` value:

```
console.assert(
  !redactedObjProxy.someFn().includes('sensitive') &&
    !redactedObjProxy.sensitiveFn().includes('sensitive') &&
    !redactedObjProxy.otherSensitiveFn().includes
      ('sensitive'),
  'JavaScript Proxy redacts correctly for all functions'
);
console.assert(
  !redactedObjProxy.field.includes('sensitive'),
  'JavaScript Proxy redacts field values by value
    correctly'
);
console.assert(
  redactedObjProxy.sensitiveField === '[REDACTED]',
```

```
    'JavaScript Proxy redacts field values by property name
      correctly'
  );
```

One of the key benefits is the simplicity of the setup; all the redaction logic is contained in the `get` function, which keeps it localized.

As an effect of the co-located logic, we've been able to add redaction by property name in addition to redacting values.

There are still some slight issues with our current Proxy-based approach since we're losing the `this` context on functions. We call `target[property](...args)`, which is fine as long as our object is not accessing `this`. We'll further refactor our implementation to make further extension easier, as well as leveraging the `Reflect` global built-in object to simplify our code.

`Reflect` provides functions with the same name as the `Proxy` trap with the same arguments; for example, `Reflect.get(target, property, receiver)`.

We'll extract a `redact` function, which takes a `propertyName` and a `redactionValue`. It will keep our redaction logic even more in sync by abstracting it to a separate function:

```
const redact = (propertyName, redactionValue) => {
  if (propertyName.includes('sensitive')) {
    return '[REDACTED]';
  }
  if (typeof redactionValue === 'string') {
    return redactionValue.replace('sensitive','[REDACTED]'
    );
  }
  // Could implement redaction of objects/Arrays and so on
  return redactionValue;
};
```

We can then use `redact` where necessary, use `Reflect.get()` as a shortcut to `target[property]`, and use `Reflect.apply` to maintain the `this` context:

```
const redactedObjProxyImproved = new Proxy(obj, {
  get(target, property, receiver) {
    const targetPropertyValue = Reflect.get(target,
      property, receiver);
    if (targetPropertyValue instanceof Function) {
      return (...args) => {
        const output = Reflect.apply(
          targetPropertyValue,
          this === receiver ? this : target,
          args
```

```
      );
        return redact(property, output);
      };
    }
    return redact(property, targetPropertyValue);
  },
});
```

Our redaction still functions the same over values, function outputs, and property and function names:

```
console.assert(
  !redactedObjProxyImproved.someFn().includes
    ('sensitive') &&
    !redactedObjProxyImproved.sensitiveFn().includes
      ('sensitive') &&
    !redactedObjProxyImproved.otherSensitiveFn().includes
      ('sensitive'),
  'JavaScript Proxy with Reflect redacts correctly for all
    functions'
);
console.assert(
  !redactedObjProxyImproved.field.includes('sensitive'),
  'JavaScript Proxy with Reflect redacts field values
    correctly'
);
console.assert(
  redactedObjProxyImproved.sensitiveField === '[REDACTED]',
  'JavaScript Proxy with Reflect redacts field values
    correctly'
);
```

Now that we've delved into how to implement the proxy pattern, we'll contrast it with the decorator pattern and which JavaScript tools we can use to implement it.

Decorator in JavaScript

The decorator pattern is similar to the proxy pattern in that it's about "wrapping" an object. However, the decorator pattern is about adding functionality to an object at runtime. Different decorators can be applied to an object to add different functionalities to it.

Implementation

Given the following `HttpClient` class based on the `fetch` API, we want to instrument the requests made through this client. `HttpClient` implements `getJson` and returns JSON output if the `fetch` request succeeds:

```
class HttpClient {
  async getJson(url) {
    const response = await fetch(url);
    if (response.ok) {
      return response.json();
    }
    throw new Error(`Error loading ${url}`);
  }
}
```

`InstrumentedHttpClient`, which is a decorator, might look like the following, where we expose the same `getJson` method but have the added `requestTimings` field on the instance.

When `getJson` is called, we track the start and end time of the `HttpClient#getJson` method call and add it to the instance's `requestTimings`:

```
class InstrumentedHttpClient {
  constructor(client) {
    this.client = client;
    this.requestTimings = {};
  }
  async getJson(url) {
    const start = performance.now();
    const output = await this.client.getJson(url);
    const end = performance.now();
    if (!Array.isArray(this.requestTimings[url])) {
      this.requestTimings[url] = [];
    }
    this.requestTimings[url].push(end - start);
    return output;
  }
}
```

We can ensure that the `InstrumentedHttpClient` works as described with the following code:

```
const httpClient = new HttpClient();
const instrumentedClient = new InstrumentedHttpClient
  (httpClient);
```

```
await instrumentedClient.getJson
  ('https://ifconfig.io/all.json');
console.assert(
  Object.keys(instrumentedClient.requestTimings).length >0,
  'Tracks request timings'
);
await instrumentedClient.getJson
  ('https://ifconfig.io/all.json');
console.assert(
  instrumentedClient.requestTimings
    ['https://ifconfig.io/all.json'].length === 2,
  'Tracks per URL timings'
);
```

Use cases

The decorator pattern, much like the proxy pattern, can be used to instrument or intercept operations on a "subject".

One key difference is that the decorator is about adding "new members" to the class, not just maintaining the interface one to one. That's why it's normal for us to save an additional requestTimings field and access it from the "decorated" class, InstrumentedHttpClient.

This means that multiple decorators can "stack" on top of each other. For example, we can have our InstrumentedHttpClient, which has requestTimings, and then create another decorator class that does something useful with the timing information. An example here is sending a "client-time" heuristic header that allows the server to stop processing a request once a certain amount time of time has passed since it knows the client will have aborted the connection by then.

Improvements/limitations

Due to JavaScript's first-class support for functions, we can use functions as the basis for decoration instead of classes.

Our getJson function could look as follows, with similar logic to the HttpClient.getJson method:

```
async function getJson(url) {
  const response = await fetch(url);
  if (response.ok) {
    return response.json();
  }
  throw new Error(`Error loading ${url}`);
}
```

We can then create an `addTiming` method that stores the request times in an `allOperationTimings` Map instance.

We're using both aspects of first-class functions here – we're passing a function as a parameter (`getJson`) and returning a function:

```
const allOperationTimings = new Map();
function addTiming(getJson) {
  return async (url) => {
    const start = performance.now();
    const output = await getJson(url);
    const end = performance.now();
    const previousOperationTimings =
        allOperationTimings.get(url) || [];
    allOperationTimings.set(url,
      previousOperationTimings.concat(end - start));
    return output;
  };
}
```

Using our decorator function is done as follows:

```
const getJsonWithTiming = addTiming(getJson);
```

We can then invoke our instrumented function and check that it adds timings to our `allOperationTimings` Map:

```
await getJsonWithTiming('https://ifconfig.io/all.json');
await getJsonWithTiming('https://ifconfig.io/all.json');
console.assert(
  allOperationTimings.size === 1,
  'operation timings tracks by url'
);
console.assert(
  allOperationTimings.get('https://ifconfig.io/all.json').
    length === 2,
  'operation timings tracks number of calls by url'
);
```

One thing you might've noticed is that our `addTiming` *is* still aware of the `getJson` interface (it knows to pass a URL parameter and that `getJson` returns a Promise object). We'll leave it as an exercise for the reader to implement, but it would be possible to turn `addTiming` into a function that can instrument the operation time of *any* JavaScript function; the tricky part is to find a good key for our operations map.

In the next part of the chapter, we'll look at the flyweight pattern.

Flyweight in JavaScript

The flyweight pattern is where the subset of object properties that have the same value are stored in shared "flyweight" objects.

The flyweight pattern is useful when generating large quantities of objects that share a subset of the same values.

Implementation

One concept from domain-driven design by Eric Evans is "value objects". These value objects have the property that their contents matter more than their identity. Let's take the example of a value object being a "coin" where, for the purposes of payment, two 50-cent coins are interchangeable.

Value objects are interchangeable and immutable (a 50-cent coin can't become a 10-cent coin). These types of objects are therefore a great fit for the Flyweight pattern.

Not all properties of a "coin" are "value"-driven, for example, certain coins are made from certain materials and coins tend to be issued in a certain year. These two properties (material and year of issue) might be interesting to collectors and in this respect, real-world coins are not only value objects as two 1993 coins might be interesting in different ways in the context of a coin collection.

We therefore model our `Wallet` as containing a list of coins and our `Coin` as containing an amount (in cents or other "minor currency"), a currency, a year of issue, and a list of materials.

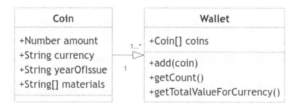

Figure 2.1: Class diagram where a Wallet has associated coins and methods to operate over them

Our `CoinFlyweight` will be our "value object" and contain the `amount` and `currency`, as follows:

```
class CoinFlyweight {
  /**
   * @param {Number} amount - amount in minor currency
   * @param {String} currency
   */
  constructor(amount, currency) {
    this.amount = amount;
```

```
        this.currency = currency;
    }
}
```

The key benefit of the flyweight pattern is that we can reuse our flyweight objects. In order to do so, we need to control the instantiation of the flyweights with a factory (as covered in *Chapter 1, Working with Creational Design Patterns*, The factory pattern in JavaScript section). We therefore define CoinFlyweightFactory with a static get method that takes the flyweight's initialization parameters but only instantiates a new CoinFlyweight if one with the right amount and currency is not already present in memory. It also provides a getCount method to return the amount of flyweights currently instantiated:

```
class CoinFlyweightFactory {
  static flyweights = {};
  static get(amount, currency) {
    const flyWeightKey = `${amount}-${currency}`;
    if (this.flyweights[flyWeightKey]) {
      return this.flyweights[flyWeightKey];
    }
    const instance = new CoinFlyweight(amount, currency);
    this.flyweights[flyWeightKey] = instance;
    return instance;
  }
  static getCount() {
    return Object.keys(this.flyweights).length;
  }
}
```

Another opportunity to use the Flyweight pattern is with materials. We can similarly create a MaterialFlyweight and reuse its values via a MaterialFlyweightFactory:

```
class MaterialFlyweight {
  constructor(materialName) {
    this.name = materialName;
  }
}

class MaterialFlyweightFactory {
  static flyweights = {};
  static get(materialName) {
    if (this.flyweights[materialName]) {
      return this.flyweights[materialName];
    }
    const instance = new MaterialFlyweight(materialName);
```

```
      this.flyweights[materialName] = instance;
      return instance;
    }
    static getCount() {
      return Object.keys(this.flyweights).length;
    }
  }
```

Finally, we can implement the `Coin` and `Wallet` classes. Our `Coin` instance has a `flyweight` field, which is populated using the `CoinFlyweightFactory`. The `Coin#materials` field is populated with a regular array but the array's contents are of `MaterialFlyweight`, loaded using the `MaterialFlyweightFactory`:

```
class Coin {
  constructor(amount, currency, yearOfIssue, materials) {
    this.flyweight = CoinFlyweightFactory.get
      (amount, currency);
    this.yearOfIssue = yearOfIssue;
    this.materials = materials.map((material) =>
      MaterialFlyweightFactory.get(material)
    );
  }
}
```

The `Wallet` is a plain JavaScript object. Its `add` method creates a new `Coin` instance and pushes it into the Wallet's `coins` field. `getTotalValueForCurrency` sums the coin's values for a given currency:

```
class Wallet {
  constructor() {
    this.coins = [];
  }
  add(amount, currency, yearOfIssue, materials) {
    const coin = new Coin(amount, currency, yearOfIssue,
      materials);
    this.coins.push(coin);
  }
  getCount() {
    return this.coins.length;
  }
  getTotalValueForCurrency(currency) {
    return this.coins
      .filter((coin) => coin.flyweight.currency ===
        currency)
```

```
      .reduce((acc, curr) => acc + curr.flyweight.amount, 0);
  }
}
```

The wallet can be used as follows, adding GBP and USD of different denominations:

```
const wallet = new Wallet();

wallet.add(100, 'GBP', '2023', ['nickel-brass',
  'nickel-plated alloy']);
wallet.add(100, 'GBP', '2022', ['nickel-brass',
  'nickel-plated alloy']);
wallet.add(100, 'GBP', '2021', ['nickel-brass',
  'nickel-plated alloy']);
wallet.add(100, 'GBP', '2021', ['nickel-brass',
  'nickel-plated alloy']);
wallet.add(200, 'GBP', '2021', ['nickel-brass',
  'cupro-nickel']);
wallet.add(100, 'USD', '1990', ['copper', 'nickel']);
wallet.add(5, 'USD', '1990', ['copper', 'nickel']);
wallet.add(1, 'USD', '2010', ['copper', 'zinc']);
```

Note that while the wallet instance contains eight coins, we've created six `CoinFlyweight` and five `MaterialFlyweight` instances:

```
console.assert(
  wallet.getCount() === 8,
  'wallet.add adds coin instances are created once given
    the same cache key'
);
console.assert(
  CoinFlyweightFactory.getCount() === 5,
  'CoinFlyweights are created once given the same
    cache key'
);
console.assert(
  MaterialFlyweightFactory.getCount() === 6,
  'MaterialFlyweights are created once given the same
    cache key'
);
console.assert(
  wallet.getTotalValueForCurrency('GBP') === 600,
  'Summing GBP works'
);
```

```
console.assert(
  wallet.getTotalValueForCurrency('USD') === 106,
  'Summing USD works'
);
```

Use cases

The flyweight pattern is a normalization technique that reduces the memory footprint at the cost of cognitive overhead when accessing and running computations over objects using this pattern. Flyweight can be leveraged as a performance optimization when handling large numbers of objects.

It's *very* well suited to modeling value objects as we've shown in the previous section. The only drawback was the getTotalValueForCurrency, where we had to read coin.flyweight.currency and coin.flyweight.amount.

Improvements/limitations

There are a few improvements we can make to our flyweight wallet/coin setup. A few of the improvements center on the "factories". The flyweights shouldn't really be accessed from outside of the get function, so we can make it a private field using #flyweights. We can also leverage the Map object, still with the same cache key, although Map has greater flexibility in terms of what keys can be used and a different property access interface (.get(key) instead of [key] access). Using a Map means we need to use this.#flyweights.size in getCount:

```
class CoinFlyweightFactory {
  static #flyweights = new Map();
  static get(amount, currency) {
    const flyWeightKey = `${amount}-${currency}`;
    if (this.#flyweights.get(flyWeightKey)) {
      return this.#flyweights.get(flyWeightKey);
    }
    const instance = new CoinFlyweight(amount, currency);
    this.#flyweights.set(flyWeightKey, instance);
    return instance;
  }
  static getCount() {
    return this.#flyweights.size;
  }
}
```

Another change we'll make is in light of the fact that there were not any material gains by making materials a flyweight, so we'll revert it to storing the list of strings per Coin instance.

Again, we want to make `#flyweight` private, this will change the interface of `Coin` since consumers will not be able to access `coin.#flyweight` (it's a private field).

What we'll do is tackle the mismatch of having to read `coin.flyweight.amount` and `coin.flyweight.currency`. We'll supply two getters, `get amount()` and `get currency()`, which will return `this.#flyweight.amount` and `this.#flyweight.currency` respectively:

```
class Coin {
  #flyweight;
  constructor(amount, currency, yearOfIssue, materials) {
    this.#flyweight = CoinFlyweightFactory.get
      (amount, currency);
    this.yearOfIssue = yearOfIssue;
    this.materials = materials;
  }
  get amount() {
    return this.#flyweight.amount;
  }
  get currency() {
    return this.#flyweight.currency;
  }
}
```

As mentioned, the interface of `Coin` doesn't have a `flyweight` property so `getTotalValueForCurrency` will read from `Coin#currency` and `Coin#amount`. As far as `Wallet` is concerned, `currency` and `amount` are fields on the `Coin` instance, although they're getters:

```
class Wallet {
  constructor() {
    this.coins = [];
  }
  add(amount, currency, yearOfIssue, materials) {
    const coin = new Coin(amount, currency, yearOfIssue,
      materials);
    this.coins.push(coin);
  }
  getCount() {
    return this.coins.length;
  }
  getTotalValueForCurrency(currency) {
    return this.coins
      .filter((coin) => coin.currency === currency)
      .reduce((acc, curr) => acc + curr.amount, 0);
```

```
    }
}
```

We can check that our new `Wallet` and `Coin` implementations work as expected by using the same tests as in our earlier iteration of the code:

```
const wallet = new Wallet();

wallet.add(100, 'GBP', '2023', ['nickel-brass',
  'nickel-plated alloy']);
wallet.add(100, 'GBP', '2022', ['nickel-brass',
  'nickel-plated alloy']);
wallet.add(100, 'GBP', '2021', ['nickel-brass',
  'nickel-plated alloy']);
wallet.add(100, 'GBP', '2021', ['nickel-brass',
  'nickel-plated alloy']);
Wallet.add(200, 'GBP', '2021', ['nickel-brass',
  'cupro-nickel']);
wallet.add(100, 'USD', '1990', ['copper', 'nickel']);
wallet.add(5, 'USD', '1990', ['copper', 'nickel']);
wallet.add(1, 'USD', '2010', ['copper', 'zinc']);

console.assert(
  wallet.getCount() === 8,
  'wallet.add adds coin instances are created once
    given the same cache key'
);
console.assert(
  CoinFlyweightFactory.getCount() === 5,
  'CoinFlyweights are created once given the same
    cache key'
);
console.assert(
  wallet.getTotalValueForCurrency('GBP') === 600,
  'Summing GBP works'
);
console.assert(
  wallet.getTotalValueForCurrency('USD') === 106,
  'Summing USD works'
);
```

We've seen how the flyweight pattern can be used to optimize memory usage by using shared value objects.

In the next part of the chapter, we'll look at the last structural design pattern covered in this book, the adapter pattern in JavaScript.

Adapter in JavaScript

The adapter pattern, similar to the other structural design patterns, focuses on interfaces.

In the adapter pattern's case, it involves being able to use a new implementation without changing the consumer or the implementation's interface. The "adapter" takes the new implementation and "adapts" the interface to match what the consumer expects.

We're not changing the implementation *or* the consumer; rather, we're building an adapter to wrap the implementation and plug it into the consumer without changing either.

Implementation

Let's start with a simple in-memory database that uses a naive `IdGenerator` to generate keys for the database entries by encoding the object as a string.

`Database` has a `createEntry` method that stores given data using the `IdGenerator` to generate a key. `Database` also has a `get` method to recall entries by ID:

```
class IdGenerator {
  get(entry) {
    return JSON.stringify(entry);
  }
}

class Database {
  constructor(idGenerator) {
    this.idGenerator = idGenerator;
    this.entries = {};
  }
  createEntry(entryData) {
    const id = this.idGenerator.get(entryData);
    this.entries[id] = entryData;
    return id;
  }
  get(id) {
    return this.entries[id];
  }
}
```

By composing `Database` with an `IdGenerator` instance, we get a key-value lookup database instance with the key equal to the JSON representation of the value:

```
const naiveIdDatabase = new Database(new IdGenerator());
naiveIdDatabase.createEntry({
  name: 'pear',
});

console.assert(
  naiveIdDatabase.get('{"name":"pear"}').name === 'pear',
  'stringIdDatabase recalls entries by stringified entry'
);
```

Now, the naive ID generation that encodes the whole entry value in the key is not ideal. An alternative is to use a UUID. Here's a `UuidFactory` using the `uuid` npm module. The key operation it exposes is `generateUuid`:

```
import { v4 as uuidv4 } from 'uuid';

class UuidFactory {
  generateUuid() {
    return uuidv4();
  }
}
```

To use the `UuidFactory` with our `Database`, we would need a `get` method instead of a `generateUuid` method. This is where our adapter comes in – we can wrap the `UuidFactory` in a class that exposes `get(entry)` but calls `generateUuid` on the `UuidFactor` instance:

```
class UuidIdGeneratorAdapter {
  constructor() {
    this.uuidFactory = new UuidFactory();
  }
  get(_entry) {
    return this.uuidFactory.generateUuid();
  }
}
```

The `UuidIdGeneratorAdapter` can then be passed as the `idGenerator` to `Database`. It all works as expected, where the entry IDs for the database are UUIDs:

```
const uuidIdDatabase = new Database(new UuidIdGeneratorAdapter());
const uuidEntryId = uuidIdDatabase.createEntry({
  name: 'pear',
});
```

```
console.assert(
  uuidIdDatabase.get(uuidEntryId).name === 'pear',
  'uuidIdDatabase recalls entries by uuid'
);
import { validate as isUuid } from 'uuid';
console.assert(isUuid(uuidEntryId), 'uuidIdDatabase generated uuid
ids');
```

Another example that makes use of the fact that the entry is being passed to idGenerator.get() is to generate prefixed auto-incrementing IDs based on the entry contents. Here, name will be used as the prefix. We have a Counter class that implements getAndIncrement(prefix), which generates incrementing IDs given a prefix (or no prefix):

```
class Counter {
  constructor(startValue = 1) {
    this.startValue = startValue;
    this.nextId = startValue;
    this.nextIdByPrefix = {};
  }
  getAndIncrement(prefix) {
    if (prefix) {
      if (!this.nextIdByPrefix[prefix]) {
        this.nextIdByPrefix[prefix] = this.startValue;
      }
      const nextId = this.nextIdByPrefix[prefix]++;
      return `${prefix}:${nextId}`;
    }
    return String(this.nextId++);
  }
}
```

Again, getAndIncrement(prefix) doesn't match the IdGenerator interface (no get method). We can wrap Counter in a PrefixedAutoIncrementIdGeneratorAdapter to expose an IdGenerator interface but using the Counter implementation:

```
class PrefixedAutoIncrementIdGeneratorAdapter {
  constructor() {
    this.counter = new Counter();
  }
  get(entry) {
    return this.counter.getAndIncrement(entry.name);
  }
}
```

We can ensure the prefixing logic works as expected for the `Database` since it creates entries keyed by prefixed auto-incrementing IDs:

```
const prefixAutoIncrementDatabase = new Database(
  new PrefixedAutoIncrementIdGeneratorAdapter()
);
```

We can check that the case where no name field is set works as expected:

```
const noPrefixIncrementingEntryId1 =
  prefixAutoIncrementDatabase.createEntry({
  type: 'no-prefix',
});
const noPrefixIncrementingEntryId2 =
  prefixAutoIncrementDatabase.createEntry({
  type: 'no-prefix',
});

console.assert(
  noPrefixIncrementingEntryId1 === '1' &&
    noPrefixIncrementingEntryId2 === '2',
  'prefixAutoIncrementDatabase generates autoincrementing
    ids with no prefix if no name property is set'
);
console.assert(
  prefixAutoIncrementDatabase.get
    (noPrefixIncrementingEntryId1).type ===
    'no-prefix' &&
    prefixAutoIncrementDatabase.get
      (noPrefixIncrementingEntryId2).type ===
      'no-prefix',
  'prefixAutoIncrementDatabase recalls entries by
    autoincrementing id'
);
```

And the scenarios where a prefix is available also functions correctly per the following example:

```
const prefixIncrementingEntryIdPear1 =
  prefixAutoIncrementDatabase.createEntry({
  name: 'pear',
});
const prefixIncrementingEntryIdPear2 =
  prefixAutoIncrementDatabase.createEntry({
  name: 'pear',
});
```

```
const prefixIncrementingEntryIdApple1 =
  prefixAutoIncrementDatabase.createEntry(
  {
    name: 'apple',
  }
);
console.assert(
  prefixIncrementingEntryIdPear1 === 'pear:1' &&
    prefixIncrementingEntryIdPear2 === 'pear:2' &&
    prefixIncrementingEntryIdApple1 === 'apple:1',
  'prefixAutoIncrementDatabase generates prefixed
    autoincrementing ids'
);
console.assert(
  prefixAutoIncrementDatabase.get
    (prefixIncrementingEntryIdPear1).name ===
    'pear',
  prefixAutoIncrementDatabase.get
    (prefixIncrementingEntryIdPear2).name ===
    'pear',
  prefixAutoIncrementDatabase.get
    (prefixIncrementingEntryIdApple1).name ===
    'apple',
  'prefixAutoIncrementDatabase recalls entries by prefixed
    id'
);
```

Use cases

The adapter pattern is useful when you need to use two classes that weren't specifically designed to work together. Consider, for example, a third-party library or module that exposes a function (such as the uuid module or even UuidFactory from our scenario). We want to abstract the implementation behind an interface, in our case the interface of IdGenerator, which is just a get method, so that any implementation can be used.

Our example showcased the value of the adapter pattern. We were able to create very differently behaving databases without changing UuidFactory, Counter, or Database for that matter. This is very important when having to connect two third-party modules or modules which are self-contained and shouldn't be changed.

Using the adapter pattern therefore means that we can avoid changing difficult-to-understand code while delivering the required functionality.

Improvements/limitations

Similarly, to the *Decorator in JavaScript - Improvements/limitations* section, one of the JavaScript features that can help when implementing structural design patterns is the first-class support for functions.

Instead of an `IdGenerator` class, we can have a `defaultIdGenerator` function that takes an entry and returns a string:

```
function defaultIdGenerator(entry) {
  return JSON.stringify(entry);
}
```

The `Database` class would now look something as follows, where `this.idGenerator(entryData)` is called directly:

```
class Database {
  constructor(idGenerator) {
    this.idGenerator = idGenerator;
    this.entries = {};
  }
  createEntry(entryData) {
    const id = this.idGenerator(entryData);
    this.entries[id] = entryData;
    return id;
  }
  get(id) {
    return this.entries[id];
  }
}
```

We can validate that the naive implementation still works by serializing whatever is passed to it as JSON:

```
const naiveIdDatabase = new Database(defaultIdGenerator);
naiveIdDatabase.createEntry({
  name: 'pear',
});

console.assert(
  naiveIdDatabase.get('{"name":"pear"}').name === 'pear',
  'stringIdDatabase recalls entries by stringified entry'
);
```

This approach shines when we need to plug in the UUID and prefix generators.

A `uuidGenerator` function can call `uuidv4()`. We can validate that `uuidIdDatabase` uses UUIDs to key and recall the entries:

```
function uuidGenerator() {
  return uuidv4();
}
const uuidIdDatabase = new Database(uuidGenerator);
const uuidEntryId = uuidIdDatabase.createEntry({
  name: 'pear',
});
console.assert(
  uuidIdDatabase.get(uuidEntryId).name === 'pear',
  'uuidIdDatabase recalls entries by uuid'
);
console.assert(isUuid(uuidEntryId), 'uuidIdDatabase
  generated uuid ids');
```

Finally, a `prefixAutoIncrementIdGenerator` would look as follows. We're using module-scoped variables, which is another feature of JavaScript:

```
const startValue = 1;
let nextId = startValue;
let nextIdByPrefix = {};
function prefixAutoIncrementIdGenerator(entry) {
  const prefix = entry.name;
  if (prefix) {
    if (!nextIdByPrefix[prefix]) {
      nextIdByPrefix[prefix] = startValue;
    }
    const nextId = nextIdByPrefix[prefix]++;
    return `${prefix}:${nextId}`;
  }
  return String(nextId++);
}
```

This code would be in a different module than its consumer, so it would be `export function prefixAutoIncrementIdGenerator` and its consumer would import `{prefixAutoIncrementIdGenerator}` from `'./path-to-module.js'`.

`prefixAutoIncrementIdGenerator` functions like the `PrefixedAutoIncrementIdGeneratorAdapter` class did, generating auto-incrementing IDs and prefixing them where possible by `entry.name`:

```
const prefixAutoIncrementDatabase = new Database(
  prefixAutoIncrementIdGenerator
);
```

```javascript
const noPrefixIncrementingEntryId1 =
  prefixAutoIncrementDatabase.createEntry({
  type: 'no-prefix',
});
const noPrefixIncrementingEntryId2 =
  prefixAutoIncrementDatabase.createEntry({
  type: 'no-prefix',
});

console.assert(
  noPrefixIncrementingEntryId1 === '1' &&
    noPrefixIncrementingEntryId2 === '2',
  'prefixAutoIncrementDatabase generates autoincrementing
    ids with no prefix if no name property is set'
);
console.assert(
  prefixAutoIncrementDatabase.get
    (noPrefixIncrementingEntryId1).type ===
    'no-prefix' &&
    prefixAutoIncrementDatabase.get
      (noPrefixIncrementingEntryId2).type ===
      'no-prefix',
  'prefixAutoIncrementDatabase recalls entries by
    autoincrementing id'
);
const prefixIncrementingEntryIdPear1 =
  prefixAutoIncrementDatabase.createEntry({
  name: 'pear',
});
const prefixIncrementingEntryIdPear2 =
  prefixAutoIncrementDatabase.createEntry({
  name: 'pear',
});
const prefixIncrementingEntryIdApple1 =
  prefixAutoIncrementDatabase.createEntry(
  {
    name: 'apple',
  }
);
console.assert(
  prefixIncrementingEntryIdPear1 === 'pear:1' &&
    prefixIncrementingEntryIdPear2 === 'pear:2' &&
    prefixIncrementingEntryIdApple1 === 'apple:1',
```

```
  'prefixAutoIncrementDatabase generates prefixed
    autoincrementing ids'
);
console.assert(
  prefixAutoIncrementDatabase.get
    (prefixIncrementingEntryIdPear1).name ===
    'pear',
  prefixAutoIncrementDatabase.get
    (prefixIncrementingEntryIdPear2).name ===
    'pear',
  prefixAutoIncrementDatabase.get
    (prefixIncrementingEntryIdApple1).name ===
    'apple',
  'prefixAutoIncrementDatabase recalls entries by prefixed
    id'
);
```

In this final section of the chapter, we covered the adapter pattern and how to use it when the consumer expects a class but also a function in JavaScript.

Summary

In this chapter, we've looked at how structural design patterns enable the extension of functionality without needing to rework interfaces in JavaScript.

The proxy design pattern is useful when we want to intercept calls to an object without changing the interface.

By contrast, the decorator design pattern concerns itself with dynamically adding functionality through new instance members.

The flyweight pattern can be used effectively for managing large numbers of objects, which is especially useful for value objects. There are workarounds in JavaScript for some of the ergonomic drawbacks of it.

The adapter pattern allows us to integrate multiple classes, modules, or functions with different opinions and interfaces without modifying them. The shape of the adapter is dictated by the existing modules and classes that we're attempting to connect together.

Now that we know how to organize relationships between different objects and classes with structural design patterns, in the next chapter, we'll cover how to use behavioral design patterns to organize communication between objects.

3

Leveraging Behavioral Design Patterns

Behavioral design patterns help to organize communication between objects. This includes the ability to extend functionality without modifying these existing classes. By implementing the behavioral design patterns covered in this chapter and how they're used in the JavaScript ecosystem, we'll learn to build JavaScript applications that can be extended without touching existing functionality.

We'll cover the following topics in this chapter:

- An understanding of the behavioral design pattern classification

- An implementation of the observer pattern and how the common Web `EventTarget` API exposes it

- Implementations of the state and strategy pattern, both with a class-based approach and a function-based approach

- A simplified visitor example, as well as common usage for the visitor pattern in the JavaScript ecosystem

By the end of this chapter, you'll be able to leverage behavioral design patterns in JavaScript to scale your code base and expose extension points for functionality.

Technical requirements

You can find the code files for this chapter on GitHub at `https://github.com/PacktPublishing/Javascript-Design-Patterns`

What are behavioral design patterns?

Communicating between objects is key to building software. Behavioral design patterns help us organize this communication and usually decouple the possible implementations from other objects. This makes us more able to extend our code base.

Behavioral design patterns help us follow the open/closed principle, where we can extend functionality without modifying the existing implementation modules.

All the patterns we'll cover allow us to "add functionality" without modifying the existing consumer/concrete implementation. In large software code bases, this is useful, since it means we can limit the scope of changes and lower the risk of breaking existing functionality. We're able to effectively de-correlate "adding functionality" from "changing the existing code for other unrelated functionality," and new features and behaviors can be added without having to do modifications to existing consumers.

With behavioral design patterns, new behaviors can be purely additive. The observer pattern allows multiple decoupled consumers (also called listeners). With the state, strategy, and visitor patterns, new implementations and transitions can be added without interfering with the existing ones.

In the next section, we'll look at our first behavioral design pattern, the observer pattern in JavaScript.

The observer pattern in JavaScript

The observer pattern allows an object (the observable or subject) to maintain a list of other objects that depend on it (observers). When a state update occurs in the subject, such as an entity object being created or updated, it notifies the observers.

Implementation

A sample use case for the observer pattern is an in-memory queue. The `Queue` instance will have the `subscribe`, `unsubscribe`, and `notify` methods.

`subscribe` will add an additional "handler" function, `unsubscribe` will remove a particular "handler" function if it has been registered, and finally, `notify` will call each handler with a "message" payload. This is the "notification of the observers" piece, where the observable or subject ensures that each registered observer is notified.

`subscribe` and `unsubscribe` turn "observer" functionality on and off, respectively. `subscribe` has to be used to become an "observer," and `unsubscribe` is useful for situations where we don't want to observe something anymore (for example, we've reached an end state). Meanwhile, the `notify` method ensures that each "subscribed" observer receives an update.

A "handler" function, as the name suggests, is a function passed to another module to be executed at that other module's discretion, usually in response to an "event":

```
class Queue {
  constructor() {
    this.handlers = [];
  }
  subscribe(handlerFn) {
    this.handlers.push(handlerFn);
  }
  unsubscribe(handlerFn) {
    this.handlers = this.handlers.filter((handler) =>
      handler !== handlerFn);
  }
  notify(message) {
    this.handlers.forEach((handler) => {
      handler(message);
    });
  }
}
```

We can implement three simple "subscribers" that will, respectively, only record 'CREATE' messages, only record 'UPDATE' messages, and record all messages:

```
const queue = new Queue();
const createMessages = [];
queue.subscribe((message) => {
  if (message.type === 'CREATE') {
    createMessages.push(message);
  }
});

const updateMessages = [];
queue.subscribe((message) => {
  if (message.type === 'UPDATE') {
    updateMessages.push(message);
  }
});

const allMessages = [];
queue.subscribe((message) => {
  allMessages.push(message);
});
```

When we trigger notifications by calling `notify`, we can ensure that the subscribers work as expected by inspecting the arrays on which they store the messages:

```
queue.notify({ type: 'CREATE', data: { user: { id: 1 } }
});
queue.notify({ type: 'CREATE', data: { user: { id: 2 } } });
queue.notify({ type: 'CREATE', data: { user: { id: 3 } } });
queue.notify({ type: 'UPDATE', data: { user: { id: 1, role:
  'ADMIN' } } });
queue.notify({
  type: 'UPDATE',
  data: { user: { id: 3, role: 'DEVELOPER' } },
});
queue.notify({ type: 'UPDATE', data: { user: { id: 3, role:
  'ADMIN' } } });

console.assert(
  createMessages.length === 3,
  '%o collects CREATE messages',
  allMessages
);
console.assert(
  updateMessages.length === 3,
  '%o collects UPDATE messages',
  allMessages
);
console.assert(
  allMessages.length === 6,
  '%o collects all message',
  allMessages
);
```

Note that our observer implementation takes advantage of first-class support for functions in JavaScript, which means we can pass a callback function to the `subscribe` method, instead of `notify` having to call a method on an instance.

In programming languages with limited or no first-class function support, such as older versions of Java and PHP, the approach would've required passing an *observer* to `subscribe` and `notify` calling a method on each observer instance. In JavaScript, if we don't use "handler" functions, we'll create an `observer` object that gets instantiated and has a `handle` function, which takes a *message* and implements some logic around it; in this case, it simply stores it on an instance variable:

```
class UpdateMessageObserver {
  constructor() {
```

```
      this.updateMessages = [];
    }
    handle(message) {
      if (message.type === 'UPDATE') {
        this.updateMessages.push(message);
      }
    }
  }
}
```

This would require modification of the Queue class to work correctly:

```
class QueueObserverObjects {
  constructor() {
    this.observers = [];
  }
  subscribe(observerObj) {
    this.observers.push(observerObj);
  }
  unsubscribe(observerObj) {
    this.observers = this.observers.filter(
      (observer) => observer !== observerObj,
    );
  }
  notify(message) {
    this.observers.forEach((observer) => {
      observer.handle(message);
    });
  }
}
```

We can ensure that it does function as expected by calling `notify` with a few messages and checking the contents of `UpdateMessageObserver().updateMessages`, as the following code sample shows:

```
const queueObserverObjects = new QueueObserverObjects();
const updateMessageObserver = new UpdateMessageObserver();
queueObserverObjects.subscribe(updateMessageObserver);
queueObserverObjects.notify({
  type: 'CREATE',
  data: { user: { id: 1 } },
});
queueObserverObjects.notify({
  type: 'UPDATE',
  data: { user: { id: 1, role: 'ADMIN' } },
```

```
});
queueObserverObjects.notify({
  type: 'UPDATE',
  data: { user: { id: 3, role: 'DEVELOPER' } },
});

console.assert(
  updateMessageObserver.updateMessages.length === 2,
  '%o collects update messages',
  updateMessageObserver.updateMessages,
);
```

We've now seen how to implement the observer pattern with "handler" functions and `Observer` object instances, with a `Queue` observable. Next, we'll look at where the observer pattern is used in JavaScript.

Use cases of the observer pattern

The observer pattern is great for dealing with loosely coupled events or messages. In the context of a web application, this could be DOM events. `EventTarget.addEventListener()` and `EventTarget.removeEventListener()`, which are available (among others) on the `Window`, `Document`, and `Element` objects, are a widely used implementation of the observer pattern. They're used by client-side JavaScript applications to register handlers for user interactions (for example, click, form submit, hover, and mouseover).

Limitations and improvements

In our queue implementation, handlers are readable from outside of the instance. The handlers are an implementation detail of the queue, which we should be able to change without affecting consuming modules. This means we want to encapsulate the handlers to make them unavailable for consumption by code outside of the `Queue` class. If we keep the `handlers` array available, it's possible for code outside of the `Queue` class to access and modify it, which means the `Queue` abstraction breaks down, since consumers integrate against implementation details. This means consumers are tightly coupled to the `Queue`'s internal implementation.

Therefore, we can use a private field; in modern JavaScript, that's done using the # syntax. For handlers, it would involve a `#handlers` declaration in the class followed by access to `this.#handlers`:

```
class Queue {
  #handlers;
  constructor() {
    this.#handlers = [];
  }
  subscribe(handlerFn) {
```

```
      this.#handlers.push(handlerFn);
    }
    unsubscribe(handlerFn) {
      this.#handlers = this.#handlers.filter((handler) =>
        handler !== handlerFn);
    }
    notify(message) {
      this.#handlers.forEach((handler) => {
        handler(message);
      });
    }
  }
```

Another improvement we can make to our queue is to provide a fluent interface so that we can "chain" calls. To do this, we simply need to return `this` from each of the `subscribe`, `unsubscribe`, and `notify` handlers. This allows us to call the instance methods in a single "chain"; instead of using `queue.subscribe()` followed by `queue.notify()`, we can write it as a single statement – `queue.subscribe().notify()`:

```
class Queue {
  #handlers;
  constructor() {
    this.#handlers = [];
  }
  subscribe(handlerFn) {
    this.#handlers.push(handlerFn);
    return this;
  }
  unsubscribe(handlerFn) {
    this.#handlers = this.#handlers.filter((handler) =>
      handler !== handlerFn);
    return this;
  }
  notify(message) {
    this.#handlers.forEach((handler) => {
      handler(message);
    });
    return this;
  }
}
```

We can validate that the queue functions as expected with regards to notifying observers, as well as being usable with the fluent ("chained") interface:

```
const queue = new Queue();
const createMessages = [];
const updateMessages = [];
const allMessages = [];

queue
  .subscribe((message) => {
    if (message.type === 'CREATE') {
      createMessages.push(message);
    }
  })
  .subscribe((message) => {
    if (message.type === 'UPDATE') {
      updateMessages.push(message);
    }
  })
  .subscribe((message) => {
    allMessages.push(message);
  });

queue
  .notify({ type: 'CREATE', data: { user: { id: 1 } } })
  .notify({ type: 'CREATE', data: { user: { id: 2 } } })
  .notify({ type: 'CREATE', data: { user: { id: 3 } } })
  .notify({ type: 'UPDATE', data: { user: { id: 1, role:
    'ADMIN' } } })
  .notify({
    type: 'UPDATE',
    data: { user: { id: 3, role: 'DEVELOPER' } },
  })
  .notify({ type: 'UPDATE', data: { user: { id: 3, role:
    'ADMIN' } } });

console.assert(
  createMessages.length === 3,
  '%o collects CREATE messages',
  allMessages
);
console.assert(
  updateMessages.length === 3,
  '%o collects UPDATE messages',
```

```
    allMessages
  );
  console.assert(
    allMessages.length === 6,
    '%o collects all message',
    allMessages
  );
```

We've now seen how to implement the observer pattern in JavaScript as well as how to use private fields and a fluent interface to improve our implementation.

In the next section, we'll implement the state and strategy patterns.

State and strategy in JavaScript and a simplified approach

The state and strategy patterns are closely related, in that they allow the extension of a software system's functionality by changing decoupled implementation objects, instead of changing the core subject object.

State allows an object to display different behavior based on what state it's in. This is very useful for modeling state machines. Each state provides the same interface, and the core object calls methods on the different states.

Strategy similarly allows an object to dynamically select an implementation at runtime. In order to do this, the implementation is injected into the object and used.

We can classify the state pattern as a subset of the strategy pattern, where the implementation is dynamically changed by the state instances.

Next, we'll see how to implement a state machine in JavaScript with the state pattern, as well as implement an object, merging abstraction with the strategy pattern.

Implementation

For our implementation of the state pattern, we'll use a simplified pull request/merge request/change request example.

A pull request starts in either a draft or open state. From there, it can transition between open and draft, and then transition to a closed or merged state. The merged state is a final state; closed can be undone by reopening the pull request, so it is not final.

To visualize the transitions from all the states, we can use a state diagram representing pull request states and allowed transitions. In *Figure 3.1*, the initial state is either draft or open. Both of these states can transition to each other. Open can change to merged or closed, where merged is a valid end state. Draft can also change to closed.

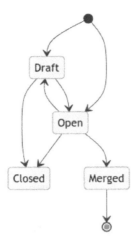

Figure 3.1: A pull request state diagram

It's useful to sketch out our `PullRequest` class first. The possible actions on our `PullRequest` are open, `markDraft`, `markReadyForReview`, close, and merge.

To implement that state pattern, we also expose a `setState` method. Each state will take the `PullRequest` instance as a constructor argument, and the initial states of `PullRequest` are either `DraftState` or `OpenState`, based on an `isDraft` boolean parameter:

```
class PullRequest {
  constructor(isDraft = false) {
    this.state = isDraft ? new DraftState(this) : new
      OpenState(this);
  }
  setState(state) {
    this.state = state;  .
  }
  open() {
    this.state.open();
  }
  markDraft() {
    this.state.markDraft();
  }
  markReadyForReview() {
    this.state.markReadyForReview();
  }
  close() {
    this.state.close();
  }
```

```
  merge() {
    this.state.merge();
  }
}
```

We'll implement the state machine, starting with the initial and final states. For the initial states, we have DraftState and OpenState; for the final states, we have MergedState.

DraftState only implements markReadyForReview and close, which transition pullRequest to OpenState or ClosedState, respectively:

```
class DraftState {
  constructor(pullRequest) {
    this.pullRequest = pullRequest;
  }
  markReadyForReview() {
    this.pullRequest.setState(new OpenState
      (this.pullRequest));
  }
  close() {
    this.pullRequest.setState(new ClosedState
      (this.pullRequest));
  }
}
```

OpenState implements markDraft, close, and merge, which transitions pullRequest to DraftState, ClosedState, and MergedState, respectively:

```
class OpenState {
  constructor(pullRequest) {
    this.pullRequest = pullRequest;
  }
  markDraft() {
    this.pullRequest.setState(new DraftState
      (this.pullRequest));
  }
  close() {
    this.pullRequest.setState(new ClosedState
      (this.pullRequest));
  }
  merge() {
    this.pullRequest.setState(new MergedState
      (this.pullRequest));
  }
}
```

As a final state, `MergedState` does not implement any of the methods:

```
class MergedState {
  constructor(pullRequest) {
    this.pullRequest = pullRequest;
  }
}
```

Finally, `ClosedState` implements the `open` method, which transitions `pullRequest` to `OpenState`:

```
class ClosedState {
  constructor(pullRequest) {
    this.pullRequest = pullRequest;
  }
  open() {
    this.pullRequest.setState(new OpenState
      (this.pullRequest));
  }
}
```

We can check that our pull request and states work as expected.

A `PullRequest` instantiated with `isDraft` set to `true` will begin in `DraftState`. A `markReadyForReview` call will transition it to `OpenState`:

```
const pullRequest1 = new PullRequest(true);
console.assert(pullRequest1.state instanceof DraftState,
  pullRequest1.state);
pullRequest1.markReadyForReview();
console.assert(pullRequest1.state instanceof OpenState,
  pullRequest1.state);
```

Once a pull request is merged with `pullRequest.merge()`, no method is available (they'll all throw errors):

```
pullRequest1.merge();

console.assert(
  captureError(() => pullRequest1.open()) instanceof Error,
  pullRequest1.state
);
console.assert(
  captureError(() => pullRequest1.markReadyForReview())
```

```
        instanceof Error,
    pullRequest1.state
  );
  console.assert(
    captureError(() => pullRequest1.close()) instanceof
      Error,
    pullRequest1.state
  );
```

A pull request starting in the open state can be closed. Once in `ClosedState`, it's not possible to do anything other than execute `open()` on it – for example, `markDraft` will fail with an error:

```
const pullRequest2 = new PullRequest(false);
console.assert(pullRequest2.state instanceof OpenState,
  pullRequest2.state);
pullRequest2.close();
console.assert(pullRequest2.state instanceof ClosedState,
  pullRequest2.state);
console.assert(
  captureError(() => pullRequest2.markDraft())
    instanceof Error,
  pullRequest2.state
);
pullRequest2.open();
console.assert(pullRequest2.state instanceof OpenState,
  pullRequest2.state);
```

We've now seen how to implement a pull request state machine using the state pattern.

Next, we'll have a look at implementing strategy.

Our example is an `ObjectMerger` class, which merges JavaScript objects. There are multiple ways to achieve this in JavaScript, so we structure our `ObjectMerger` to accept a `strategy` object and allow updates to it with a `setStrategy` method. Finally, we expose a `combinedObjects` method, which calls the instance's strategy's `combineObjects` method with two objects:

```
class ObjectMerger {
  constructor(defaultStrategy) {
    this.strategy = defaultStrategy;
  }
  setStrategy(newStrategy) {
    this.strategy = newStrategy;
  }
```

```
  combineObjects(obj1, obj2) {
    return this.strategy.combineObjects(obj1, obj2);
  }
}
```

An example strategy in this case would be to use `Object.assign` with `{}` (a new object literal) as the target of the assignment. This has the benefit of not mutating the `obj1` and `obj2` parameters:

```
class PureObjectAssignStrategy {
  constructor() {}
  combineObjects(obj1, obj2) {
    return Object.assign({}, obj1, obj2);
  }
}
```

Our `ObjectMerger` can be instantiated with the `PureObjectAssignStrategy`, as follows:

```
const objectMerger = new ObjectMerger
  (new PureObjectAssignStrategy());
```

It can then be used to merge objects without mutating `obj1` or `obj2`:

```
const obj1 = {
  keys: '123',
};
const obj2 = {
  keys: '456',
};

const defaultMergeStrategyOutput =
  objectMerger.combineObjects(obj1, obj2);
console.assert(defaultMergeStrategyOutput.keys === '456',
  '%o has keys = 456');
console.assert(obj1.keys === '123' && obj2.keys === '456',
  obj1, obj2);
```

An example of a naive implementation using `Object.assign` that doesn't use a new object as the assignment target (and, therefore, mutates `obj1`) looks as follows:

```
class MutatingObjectAssignStrategy {
  constructor() {}
  combineObjects(obj1, obj2) {
    return Object.assign(obj1, obj2);
  }
}
```

It can be used as follows and does indeed mutate `obj1`:

```
objectMerger.setStrategy(new
  MutatingObjectAssignStrategy());
const mutatingMergedStrategyOutput =
  objectMerger.combineObjects(obj1, obj2);
console.assert(
  mutatingMergedStrategyOutput.keys === '456',
  '%o has keys = 456',
  mutatingMergedStrategyOutput
);
console.assert(
  obj1.keys === '456' && obj2.keys === '456',
  'Mutates the original object obj1 %o, obj2 %o',
  obj1,
  obj2
);
```

An equivalent strategy to our initial `Object.assign({}, obj1, obj2)` strategy is to use the spread syntax:

```
class ObjectSpreadStrategy {
  constructor() {}
  combineObjects(obj1, obj2) {
    return { ...obj1, ...obj2 };
  }
}
```

We can validate that spreading `obj1` and `obj2` yields the same strategy characteristics as our earlier `PureObjectAssignStrategy`:

```
objectMerger.setStrategy(new ObjectSpreadStrategy());

const newObj1 = { keys: '123' };
const newObj2 = { keys: '456', obj1: newObj1 };

const objectSpreadStrategyOutput =
  objectMerger.combineObjects(
  newObj1,
  newObj2
);
console.assert(
  objectSpreadStrategyOutput.keys === '456',
  '%o has keys = 456',
  objectSpreadStrategyOutput
```

```
);
console.assert(
  newObj1.keys === '123' && newObj2.keys === '456',
  'Does not mutate the original object newObj1 %o,
    newObj2 %o',
  newObj1,
  newObj2
);
```

One interesting aspect is that this approach only creates a shallow clone; object references inside of the objects are copied, but the contents of the target objects are the same:

```
console.assert(
  objectSpreadStrategyOutput.obj1 === newObj1,
  'Does a shallow clone so objectSpreadStrategyOutput.obj1
    references newObj1'
);
```

We can remediate this by implementing a deep cloning strategy based on `structuredClone`:

```
class DeepCloneObjectAssignStrategy {
  constructor() {}
  combineObjects(obj1, obj2) {
    return Object.assign(structuredClone(obj1),
      structuredClone(obj2));
  }
}
```

`DeepCloneObjectAssignStrategy` has all the properties of `PureObjectAssignStrategy` and `ObjectSpreadStrategy`, with the addition of doing a deep copy, recursively copying the contents of nested objects instead of copying references to those objects:

```
objectMerger.setStrategy(new DeepCloneObjectAssignStrategy());

const deepCloneStrategyOutput = objectMerger.
  combineObjects(newObj1, newObj2);
console.assert(
  deepCloneStrategyOutput.keys === '456',
  '%o has keys = 456',
  deepCloneStrategyOutput
);
console.assert(
  newObj1.keys === '123' && newObj2.keys === '456',
  'Does not mutate the original object newObj1 %o,
    newObj2 %o',
```

```
    newObj1,
    newObj2
  );

  console.assert(
    deepCloneStrategyOutput.obj1 !== newObj1 &&
      deepCloneStrategyOutput.obj1.keys === newObj1.keys,
    'Does a shallow clone so deepCloneStrategyOutput.obj1
      references newObj1'
  );
```

We've now seen how to implement the state and strategy patterns. Next, we'll look at where the state and strategy patterns are most often used in JavaScript.

Use cases of the state and strategy patterns

As mentioned earlier in the chapter, the state pattern is useful for implementing state machines.

A key difference between state and strategy is that, in the state pattern, it tends to be the case that different states know about each other – for example, `ClosedState` creates a new instance of `OpenState` to transition to it. Similarly, `OpenState` is aware of all the potential states it can be transitioned to (`DraftState`, `ClosedState`, and `MergedState`). In contrast, when implementing the strategy pattern, different strategies are self-contained and not aware of each other. For example, `PureObjectAssignStrategy` and `MutatingObjectAssignStrategy` don't reference each other.

Strategy is useful to provide a consistent interface with different internal implementations. It's a useful abstraction when different implementing algorithms should be swappable without an integrating consumer knowing about it.

Limitations and improvements

In our state example, note how much of our code is a duplicated class constructor, which takes a `pullRequest` instance. We can refactor our code by providing a `PullRequestBaseState` class, which throws `IllegalOperationError` for each of the state methods:

```
  class IllegalOperationError extends Error {
    constructor(stateInstance) {
      this.stateInstance = stateInstance;
      throw new Error('Illegal operation for State');
    }
  }
  class PullRequestBaseState {
    constructor(pullRequest) {
```

```
      this.pullRequest = pullRequest;
    }
    markDraft() {
      throw new IllegalOperationError(this);
    }
    markReadyForReview() {
      throw new IllegalOperationError(this);
    }
    open() {
      throw new IllegalOperationError(this);
    }
    close() {
      throw new IllegalOperationError(this);
    }
    merge() {
      throw new IllegalOperationError(this);
    }
  }
```

This means we can define our different states by extending `PullRequestBaseState`:

```
  class ClosedState extends PullRequestBaseState {
    open() {
      this.pullRequest.setState(new OpenState
        (this.pullRequest));
    }
  }

  class DraftState extends PullRequestBaseState {
    markReadyForReview() {
      this.pullRequest.setState(new OpenState
        (this.pullRequest));
    }
    close() {
      this.pullRequest.setState(new ClosedState
        (this.pullRequest));
    }
  }

  class OpenState extends PullRequestBaseState {
    markDraft() {
      this.pullRequest.setState(new DraftState
        (this.pullRequest));
    }
  }
```

```
  close() {
    this.pullRequest.setState(new ClosedState
      (this.pullRequest));
  }
  merge() {
    this.pullRequest.setState(new MergedState
      (this.pullRequest));
  }
}
class MergedState extends PullRequestBaseState {}
```

The `PullRequest` class doesn't change, and these new state implementations work the same as our previous implementation:

```
const pullRequest1 = new PullRequest(true);
console.assert(pullRequest1.state instanceof DraftState,
  pullRequest1.state);
pullRequest1.markReadyForReview();
console.assert(pullRequest1.state instanceof OpenState,
  pullRequest1.state);
pullRequest1.merge();
console.assert(
  captureError(() => pullRequest1.open()) instanceof Error,
  pullRequest1.state
);
console.assert(
  captureError(() => pullRequest1.markReadyForReview())
    instanceof Error,
  pullRequest1.state
);
console.assert(
  captureError(() => pullRequest1.close()) instanceof
    Error,
  pullRequest1.state
);

const pullRequest2 = new PullRequest(false);
console.assert(pullRequest2.state instanceof OpenState,
  pullRequest2.state);
pullRequest2.close();
console.assert(pullRequest2.state instanceof ClosedState,
  pullRequest2.state);
console.assert(
  captureError(() => pullRequest2.markDraft())
```

```
      instanceof Error,
    pullRequest2.state
  );
  pullRequest2.open();
  console.assert(pullRequest2.state instanceof OpenState,
    pullRequest2.state);
```

For strategy, one thing we can leverage is JavaScript's first-class function support. Instead of implementing each strategy as an object, we can make them functions.

Our `ObjectMerger`'s implementation looks as follows:

```
class ObjectMerger {
  constructor(defaultStrategy) {
    this.strategy = defaultStrategy;
  }
  setStrategy(newStrategy) {
    this.strategy = newStrategy;
  }
  combineObjects(obj1, obj2) {
    return this.strategy(obj1, obj2);
  }
}
```

We can then re-implement all our strategies as functions:

```
function pureObjectAssignStrategy(obj1, obj2) {
  return Object.assign({}, obj1, obj2);
}
function mutatingObjectAssignStrategy(obj1, obj2) {
  return Object.assign(obj1, obj2);
}
function objectSpreadStrategy(obj1, obj2) {
  return { ...obj1, ...obj2 };
}
function deepCloneObjectAssignStrategy(obj1, obj2) {
  return Object.assign(structuredClone(obj1),
    structuredClone(obj2));
}
```

The function-based strategy `ObjectMerger` class has the same attributes as the class-based one that we implemented earlier. The constructor takes a "strategy function", which it sets on the instance; each instance exposes a `setStrategy` method, which overrides the strategy function, and a `combineObjects` method, which we can call to merge objects.

This means we can use our `ObjectMerger` with all four function-based strategies (`pureObjectAssignStrategy`, `mutatingObjectAssignStrategy`, `objectSpreadStrategy`, and `deepCloneObjectAssignStrategy`), as the following demonstrates:

```
const objectMerger = new ObjectMerger
  (pureObjectAssignStrategy);
const obj1 = {
  keys: '123',
};
const obj2 = {
  keys: '456',
};

const defaultMergeStrategyOutput =
  objectMerger.combineObjects(obj1, obj2);
console.assert(defaultMergeStrategyOutput.keys === '456',
  '%o has keys = 456');
console.assert(obj1.keys === '123' && obj2.keys === '456',
  obj1, obj2);
objectMerger.setStrategy(mutatingObjectAssignStrategy);
const mutatingMergedStrategyOutput =
  objectMerger.combineObjects(obj1, obj2);
console.assert(
  mutatingMergedStrategyOutput.keys === '456',
  '%o has keys = 456',
  mutatingMergedStrategyOutput
);
console.assert(
  obj1.keys === '456' && obj2.keys === '456',
  'Mutates the original object obj1 %o, obj2 %o',
  obj1,
  obj2
);

objectMerger.setStrategy(objectSpreadStrategy);

const newObj1 = { keys: '123' };
const newObj2 = { keys: '456', obj1: newObj1 };

const objectSpreadStrategyOutput =
  objectMerger.combineObjects(
  newObj1,
  newObj2
```

```
);
console.assert(
  objectSpreadStrategyOutput.keys === '456',
  '%o has keys = 456',
  objectSpreadStrategyOutput
);
console.assert(
  newObj1.keys === '123' && newObj2.keys === '456',
  'Does not mutate the original object newObj1 %o,
    newObj2 %o',
  newObj1,
  newObj2
);
console.assert(
  objectSpreadStrategyOutput.obj1 === newObj1,
  'Does a shallow clone so objectSpreadStrategyOutput.obj1
    references newObj1'
);
objectMerger.setStrategy(deepCloneObjectAssignStrategy);
const deepCloneStrategyOutput = objectMerger.combineObjects
  (newObj1, newObj2);
console.assert(
  deepCloneStrategyOutput.keys === '456',
  '%o has keys = 456',
  deepCloneStrategyOutput
);
console.assert(
  newObj1.keys === '123' && newObj2.keys === '456',
  'Does not mutate the original object newObj1 %o,
    newObj2 %o',
  newObj1,
  newObj2
);

console.assert(
  deepCloneStrategyOutput.obj1 !== newObj1 &&
    deepCloneStrategyOutput.obj1.keys === newObj1.keys,
  'Does a shallow clone so deepCloneStrategyOutput.
    obj1 references newObj1'
);
```

We've shown how to implement the state and strategy patterns in JavaScript, as well as their limitations and improvements, which can be done using modern JavaScript features.

In the next section, we'll introduce the visitor pattern and its usage in the JavaScript ecosystem.

Visitor in JavaScript

The visitor design pattern concerns itself with being able to add functionality to objects without modifying the structure of them.

With classical inheritance, we often end up with a "base class" that is not used directly; it's used as an "abstract class," from which "concrete" classes inherit from our "base class." For example, with BankAccount and BankAccountWithInterest, our class diagram would look as follows, where BankAccountWithInterest extends BankAccount and overrides setBalance.

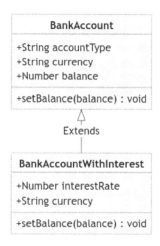

Figure 3.2: A class diagram for BankAccountWithInterest inheriting from BankAccount

What we can do with the visitor pattern is define BankAccount, which accepts a visitor and an InterestRateVisitor visitor class. As a class diagram, it looks as follows. BankAccount and InterestRateVisitor are not linked via inheritance; they will be linked at runtime when InterestRateVisitor is called by the BankAccount().accept method. This means InterestRateVisitor knows about the structure of BankAccount but not the other way around. Furthermore, a visitor might not need to know the full structure of what it's visiting, only what's relevant to implement the visitor's functionality.

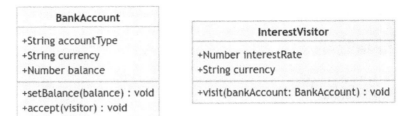

Figure 3.3: A class diagram for BankAccount and an independent InterestRateVisitor

We'll now see how to implement the `BankAccount` and `InterestRateVisitor` scenario.

Implementation

To implement a visitor, let's start with a simple `BankAccount` class. The constructor sets the account type (either a current account or a savings account), the currency, and the initial balance. `BankAccount` has a `setBalance` method that can set the value of an account's balance. The `accept` method will allow us to accept visitors and call their `visit` method on the instance:

```
class BankAccount {
  /**
   *
   * @param {'CURRENT' | 'SAVINGS'} accountType
   * @param {String} currency
   * @param {Number} balance - balance in minor currency
     unit
   */
  constructor(accountType = 'CURRENT', currency = 'USD',
    balance = 0) {
    this.accountType = accountType;
    this.currency = currency;
    this.balance = balance;
  }
  setBalance(balance) {
    this.balance = balance;
  }
  accept(visitor) {
    visitor.visit(this);
  }
}
```

A way to structure `InterestVisitor` is to initialize it with an interest rate and currency. The `visit` method takes `bankAccount` and, if the account matches the currency and is a savings account, applies a new balance, based on the interest rate and current balance:

```javascript
class InterestVisitor {
  constructor(interestRate, currency) {
    this.interestRate = interestRate;
    this.currency = currency;
  }
  /**
   * @param {BankAccount} bankAccount
   */
  visit(bankAccount) {
    if (
      bankAccount.currency === this.currency &&
      bankAccount.accountType === 'SAVINGS'
    ) {
      bankAccount.setBalance((bankAccount.balance *
        this.interestRate) / 100);
    }
  }
}
```

Given a set of accounts, we can create USD and GBP `InterestVisitor` instances:

```javascript
const accounts = [
  new BankAccount('SAVINGS', 'GBP', 500),
  new BankAccount('SAVINGS', 'USD', 500),
  new BankAccount('CURRENT', 'USD', 10000),
];

const usdInterestVisitor = new InterestVisitor(105, 'USD');
const gbpInterestVisitor = new InterestVisitor(110, 'GBP');
```

We can then loop through the accounts and call the `accept` method with the relevant visitor:

```javascript
accounts.forEach((account) => {
  account.accept(usdInterestVisitor);
  account.accept(gbpInterestVisitor);
});

console.assert(
  accounts[0].balance === 550 &&
    accounts[1].balance === 525 &&
    accounts[2].balance === 10000,
```

```
'%o',
accounts
);
```

We've now seen how to implement the visitor pattern in a band account scenario. Next, we'll look at popular use cases for the visitor pattern in JavaScript.

Use cases of the visitor pattern

The visitor pattern provides a simple interface for library authors to allow consumers to extend a library's functionality. This is especially effective in libraries that deal with trees or other "sets of nodes". This explains why the visitor pattern is popular for custom plugins for parsing systems such as GraphQL implementations, or compilers such as Babel.

For example, the way to write a custom directive in Apollo Server v2 is to extend `SchemaDirectiveVisitor`:

```
import { SchemaDirectiveVisitor } from 'apollo-server';
class CustomDirective extends SchemaDirectiveVisitor {
  visitFieldDefinition(field) {
    // we can replace/augment the field's resolver
        implementation here
  }
}
```

The Babel compiler can be extended using a visitor. For example, the following visitor inspects function declaration names for a given code snippet, parsed using the `'@babel/parser'` package:

```
import * as parser from '@babel/parser';
import traverse from '@babel/traverse';

const ast = parser.parse(`function triple(n) {
  return n * 3;
}`);

const CustomVisitor = {
  FunctionDeclaration(path) {
    console.assert(path.node.id.name === 'triple');
  },
};

traverse(ast, CustomVisitor);
```

We've now seen how the visitor pattern is used for libraries that manipulate tree data structures. Next, we'll recapitulate what we've learned in this chapter.

Summary

In this chapter, we saw how behavioral design patterns enable the extension of functionality by supporting different implementations and decoupling parts of the code base.

The observer pattern is useful to support communication with loosely coupled observable/observer pairs. The state and strategy patterns can be used to implement state machines and swap implementations effectively. The visitor pattern is a great way to expose an extension mechanism that's decoupled from the structure of the objects it's operating on.

Now that we know how to organize communication between different objects and classes with behavioral design patterns, in the next chapter, we'll cover reactive view library patterns in React.

Part 2: Architecture and UI Patterns

In this part, you will get an overview of architecture and UI patterns in JavaScript. You will learn about common reactive view library patterns in React and rendering strategies with React and Next. js. Finally, you'll learn about two approaches to scaling your application via micro frontends with the **zones** and **islands** architectures.

This part has the following chapters:

- *Chapter 4, Exploring Reactive View Library Patterns*
- *Chapter 5, Rendering Strategies and Page Hydration*
- *Chapter 6, Micro Frontends, Zones, and Islands Architectures*

4

Exploring Reactive View Library Patterns

Reactive view library patterns give us tools to build applications in a scalable and maintainable manner when we could benefit from breaking out of the component primitive. Using the React view library, we'll cover different techniques for going beyond component-based composition to inject functionality into our components – the render prop, a higher-order component, hooks, and provider patterns.

We'll cover the following main topics in this chapter:

- An introduction to reactive view library patterns and where we can benefit the most by using them
- Examples and implementation approaches for the render prop pattern
- Implementing and using the higher-order component pattern
- Using hooks to build React function components
- Multiple ways to implement the provider pattern

By the end of this chapter, you'll be able to discern when and how to use Reactive view library patterns to build React applications.

Technical requirements

To follow along in this chapter, you'll need the following:

- **Node.js 20**+: `https://nodejs.org/en`
- **Npm 8**+: Comes with most Node.js installations

- `https://parceljs.org/` is used in some examples and has similar platform support to Node.js

- **React**: React DOM and Formik are installed via npm; an understanding of `https://react.dev/` basics in a web context is required

You can find the code files for this chapter on GitHub at `https://github.com/PacktPublishing/Javascript-Design-Patterns`

What are reactive view library patterns?

Reactive view libraries are extensively used for JavaScript and web frontend development. A few very popular options are **React**, **Angular**, and **Vue**.

Reactive view libraries provide a way to write applications in a more scalable fashion by allowing the user interface (usually the browser) to react to changes in the data. Application development is, therefore, simplified, since the view library or framework takes care of all the direct manipulation necessary to maintain synchronization between the underlying data and the browser.

One of the key common denominators between these libraries and frameworks is the concept of a **component**, which contains business logic and/or rendering logic. The component is a key building block of an application. It can be reused or not, but it usually encapsulates a set of responsibilities and enforces interfaces around it.

A trait of components is that a developer should be able to use them as building blocks, and without component internals changing significantly, an application's behavior can be changed significantly.

Reactive view library patterns, therefore, help us build components in a reusable fashion, but they also cover techniques to work around situations where the component abstraction has shortcomings.

In the following sections, we'll cover the render props, higher-order component, hooks, and the provider pattern in React. We'll be focusing on React, but the patterns have equivalents in Vue.

The render prop pattern

The render prop pattern is apparent when a component allows its consumer to define how a part of that component is rendered, via a function prop. These can be children as a function or another prop, which is a function that takes some parameters and returns JSX.

Render props allow for a level of inversion of control. Although a component could completely encapsulate rendering and business logic, it instead yields control of some parts of the rendering logic to its consumer.

Such inversion of control is useful to share logic without sharing the visuals or actually rendering the UI. Therefore, this pattern is widespread among libraries. A prime example is **Formik**, which gives consumers flexibility on how to render a form while providing an abstraction over the form's state management logic.

Use cases

Let's start with a scenario where we build a `CoupledSelect` component, which is a wrapper for the `select` native element. We'll build this component in a way that the data and the rendering are closely coupled, providing a simple example of when render props can be useful.

A consumer's expectation from `CoupledSelect` is that it would behave mostly like the `select` native element, with some caveats.

Our `CoupledSelect` component takes the following props:

- `selectedOption`: This sets the selected option; it is akin to the selected prop on the `option` native element
- `options`: This is an array of strings that are rendered as `option` elements
- `onChange`: This is an optional callback for the component rendering `CoupledSelect` to react to option selections

We can implement it as follows. `CoupledSelect` will wrap around `onChange`, since it's optional:

```
import React from 'react';
export function CoupledSelect({ selectedOption, options,
  onChange }) {
  const onChangeHandler = (event) => {
    if (onChange) onChange(event.target.value);
  };
}
```

Let's move on to the rendering logic. We'll return a select element with onChange={onChangeHandler} and value={selectedOption} so that select will be in sync with selectedOption and propagate changes back:

```
import React from 'react';
export function CoupledSelect({ selectedOption, options,
  onChange }) {
  const onChangeHandler = (event) => {
    if (onChange) onChange(event.target.value);
  };
  return <select onChange={onChangeHandler}
    value={selectedOption}></select>;
}
```

Finally, we'll render the `props.options` using .map, which will return an `<option>` element, with the value and key properties set to option and whose content will be the `option` value also:

```
export function CoupledSelect({ selectedOption, options,
  onChange }) {
  // no change to onChangeHandler
  return (
    <select onChange={onChangeHandler}
      value={selectedOption}>
      {options.map((option) => (
        <option value={option} key={option}>
          {option}
        </option>
      ))}
    </select>
  );
}
```

Using our `CoupledSelect` might look something as follows.

We define an array of options. Here, we structure them as a list of objects with a `value` key that's a string:

```
const options = [
  { value: 'apple' },
  { value: 'pear' },
  { value: 'orange' },
  { value: 'grape' },
  { value: 'banana' },
];
```

We can then use `CoupledSelect` by ensuring that `props.options` is an array of strings:

```
function App() {
  return (
    <>
      <CoupledSelect
        options={options.map((option) => option.value)}
      />
    <>
  );
}
```

Next, we can save `selectedOption` by using the `useState` hook. We will name this particular piece of state `selectedOption` and its update function `setSelectedOption`. This will allow us to make `CoupledSelect` interactive:

```
function App() {
  const [selectedOption, setSelectedOption] = useState();
  return (
    <>
      <p>Selected Option: {selectedOption}</p>
      <CoupledSelect
        selectedOption={selectedOption}
        onChange={(selectedOption) => setSelectedOption
          (selectedOption)}
        options={options.map((option) => option.value)}
      />
    <>
  );
}
```

Finally, we will set an initial value for `selectedOption` to show that this functionality of the `CoupledSelect` component works:

```
function App() {
  const [selectedOption, setSelectedOption] = useState
    (options[3].value);
  // no change to the returned JSX
}
```

Starting with the initial `selectedOption` functionality, we can see that the item at index 3 of the options array, `{ value: 'grape' }`, is the initially selected option, as shown in *Figure 4.1*:

Selected Option: grape

Figure 4.1: The CoupledSelect initial state, with grape selected

When opening `select`, the **grape** is also selected, which means `select` is in the correct state.

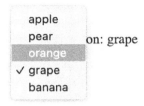

Figure 4.2: The CoupledSelect select open state, hovering on the orange option

Finally, when we select a different option, **orange**, we see it's reflected both in the select element and out of it (see *Figure 4.3*), meaning our integrated `onChange` handler also works as expected.

Selected Option: orange

Figure 4.3: In the CoupledSelect post-selection state, orange is now selected

The `CoupledSelect` component has limited flexibility due to the `options.map()` call in the render function. Since we're using the option variable as the option element's value, it must be a string or number. The value is also equal to the rendered text content of the option element, but there are often situations where we want to display something different from the value that we're storing. It's a presentation versus persistence concern here. For example, we can't change the values rendered without changing what's being stored in `onChange`.

If we want to add a `Fruit:` prefix to `select`, a naive approach is to implement it as follows:

```
<CoupledSelect
  {/* other props don't change */}
  options={options.map((option) => `Fruit:
    ${option.value}`)}
/>
```

Unfortunately, this doesn't work as expected, as the initial selection doesn't work anymore:

Selected Option: orange

Fruit: apple   ∨

Figure 4.4: The CoupledSelect initial state, where the initial selection does not work correctly

When opening the `select`, things seem to work OK; we can see the **Fruit:** prefix for all the options, as shown in *Figure 4.5*.

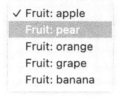

Figure 4.5: The CoupledSelect open state, with the Fruit: prefix

On selection of a new option, we can see that what's being stored in the `selectedOption` is **Fruit: pear** instead of **pear**.

Selected Option: Fruit: pear

Figure 4.6: The CoupledSelect post-selection state – note that the
selected option Fruit: pear includes the prefix Fruit:

Therefore, the `CoupledSelect` component can't be used flexibly, due to the coupling of the render logic and the data logic.

We'll now see how render props could have alleviated this issue by decoupling data and rendering logic.

Implementation/example

In our `CoupledSelect` example, we saw how the stored data and what is displayed to the user are tightly coupled. We'll now see how to break that coupling using render props.

Decoupling presentation from data logic by using render props

An alternative way to have written the `CoupledSelect` component with render props is as follows. The key additional prop we're passing is `renderOption`, a render prop. Most of the remaining components are similar but are included for completeness:

```
export function SelectRenderProps({
   selectedOption,
   options,
   renderOption,
   onChange,
}) {
```

```
    const onChangeHandler = (event) => {
      if (onChange) onChange(event.target.value);
    };
    return (
      <select onChange={onChangeHandler} value=
        {selectedOption}>
        {options.map((option) => renderOption(option))}
      </select>
    );
  }
```

Usage of the `SelectRenderProps` component is very similar to that of `CoupledSelect`. The only additional prop we need to provide an implementation for is the `renderOption` prop, which we implement with a function that returns an `option` element:

```
function App() {
  return (
    <SelectRenderProps
      selectedOption={selectedOption}
      onChange={(selectedOption) => setSelectedOption
        (selectedOption)}
      options={options.map((option) => option.value)}
      renderOption={(option) => (
        <option value={option} key={option}>
          {option}
        </option>
      )}
    />
  );
}
```

So far, the implementation is very similar to `CoupledSelect`, apart from the fact that the parent of `SelectRenderProps` now decides how to render an option.

Given the same requirement to prefix the options with `Fruit:`, we can now implement it as follows:

```
<SelectRenderProps
  {/* rest of the props remain unchanged */}
  renderOption={(option) => (
    <option value={option} key={option}>
      Fruit: {option}
    </option>
  )}
/>
```

Note that counter to what we did with `CoupledSelect`, we're not even touching the options prop. Our only change is to the `renderOption` prop. We'll now test this example and show that decoupling the rendering logic (with a render prop) from the data logic works much better for extensibility.

The `SelectRenderProps` initial state renders correctly, with **Fruit: grape** inside `select` and **grape** in the parent component:

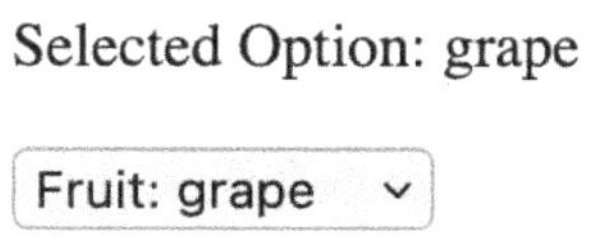

Selected Option: grape

Fruit: grape ⌄

Figure 4.7: The SelectRenderProps initial state – the options and initial selection are displayed correctly

When we open `select`, we can see that the **Fruit:** prefix is rendered.

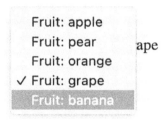

Figure 4.8: The SelectRenderProps open state – the options include the Fruit: prefix

Finally, upon selection of an option, the state is updated correctly, the parent stores **banana**, and `select` has **Fruit: banana** selected:

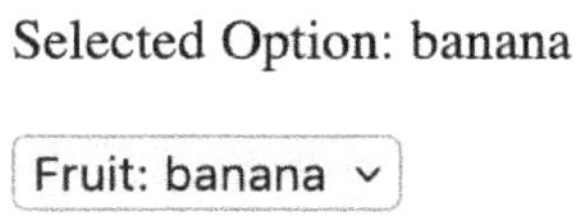

Selected Option: banana

Fruit: banana ⌄

Figure 4.9: The SelectRenderProps post-selection state – the
selected option does not include the Fruit: prefix

We've now seen how render props can allow the rendering logic and data logic to be edited separately when making a rendering change.

Now that we've implemented a basic example of the render prop pattern, we'll see how libraries leverage it to provide flexibility to consumers.

Additional render prop patterns when providing components with flexible presentation

The React form management library Formik uses a render prop to provide form state back to consumers. The render prop is the children prop of the Formik component. In other words, what's between the opening `<Formik>` tag and the closing `</Formik>` tag is a function, which provides props such as values, `isSubmitting`, and `handleChange`.

See the following example, which is a single-input form that takes a name, validates that it's at least two characters long, and allows the form to be submitted.

To begin, we'll render the form and input that will store the `fields` value in `Formik`:

```
import { Formik } from 'formik';

export function FormikIntegrationExample() {
  return (
    <Formik
      initialValues={{ name: '' }}
    >
      {(({
        values,
        errors,
        touched,
        handleChange,
        handleBlur,
        handleSubmit,
        isSubmitting,
      }) => (
        <form onSubmit={handleSubmit}>
          <fieldset>
            <input
              type="text"
              id="name"
              name="name"
              onChange={handleChange}
              onBlur={handleBlur}
              value={values.name}
              aria-required="true"
            />
          </fieldset>
        </form>
      )}
    </Formik>
```

```
  );
}
```

Next, we can add submission handling and an inline validation error display:

```
import { Formik } from 'formik';

export function FormikIntegrationExample() {
  return (
    <Formik
      initialValues={{ name: '' }}
      validate={(values) => {
        const errors = {};

        if (!values.name) {
          errors.name = 'Required';
        } else if (values.name.length < 2) {
          errors.name = 'Name too short';
        }

        return errors;
      }}
      onSubmit={(values, { setSubmitting }) => {
        setTimeout(() => {
          alert(JSON.stringify(values, null, 2));

          setSubmitting(false);
        }, 400);
      }}
    >
      {({
        /* no change to props in render prop */
      }) => (
        <form onSubmit={handleSubmit}>
          <fieldset>
            <div>
              <label htmlFor="name">
                Name (Required)
                <br />
                {errors.name && touched.name ? (
                  <>Error: {errors.name}</>
                ) : (
                  <> </>
                )}
```

```
            </label>
          </div>
          {/* no change to the input */}
        </fieldset>

        <button type="submit" disabled={isSubmitting}>
          Submit
        </button>
      </form>
    )}
  </Formik>
);
}
```

In the initial state, we see the form with a single input and a submit button:

Figure 4.10: The Formik single field and Submit button in its initial
state, which includes the Name (Required) label

When we click (or otherwise focus) on the name input and then un-focus (the *blur* web event), the
validation triggers, letting us know that the field is required.

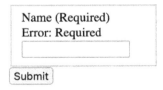

Figure 4.11: Name input blur validation - Error: Required validation error

If we input only one character and blur, we get a validation error, **Name too short**.

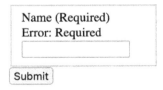

Figure 4.12: H in the name input triggers the validation error Name too short

When a name that meets the validation criteria is met, the validation errors are cleared.

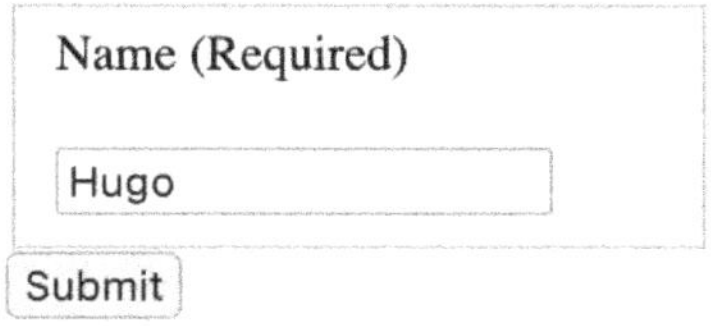

Name (Required)
Hugo
Submit

Figure 4.13: A valid Formik field clears the validation errors

Finally, when we click **Submit**, we get a browser alert with { **"name": "Hugo"** }.

Name (Required)
Hugo
localhost:1234
{
"name": "Hugo"
}
OK

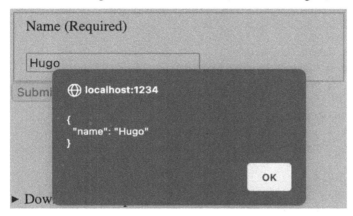

Figure 4.14: An alert on submission with {"name": "Hugo"}

Now, let's take a look at some limitations of the render prop pattern.

Limitations

One key limitation of the render prop pattern is that it provides units of reuse and integration that are functions and not components. It's possible for a lot of the logic to end up in the render prop function itself that could have been better served by creating a new component.

Render props can make code harder to test when using a shallow renderer such as Enzyme's `shallow`, which won't render the full component tree. Components making heavy use of render props should probably use a full "mount" rendering approach so that all the children of the component (including the render props) are rendered.

In this section, we introduced you to the render prop pattern and described its use cases, examples, and limitations.

In the next section, we'll learn about another reactive view library pattern – higher-order components.

The higher-order component pattern

A higher-order component is a function that takes a component and returns a component. The definition of higher-order components is similar to higher-order functions, which JavaScript supports. Higher-order functions are functions that receive a function as a parameter or return a function.

The higher-order component pattern allows us to pass additional props to a component.

Implementation/example

The following is a simple render prop, `withLocation`, which injects `window.location.href` and `window.location.origin` into a component as props:

```
const location = {
  href: window.location.href,
  origin: window.location.origin,
};

export function withLocation(Component) {
  return (props) => {
    return <Component location={location} {...props} />;
  };
}
```

The pattern that's used when using higher-order components is to export default the higher-order component called with the local component – in this case, `withLocation(Location)`. The `Location` component is a simple component that takes location as provided by `withLocation` and renders it:

```
// in `location.jsx` file
function Location({ location }) {
  return (
    <>
      location.href: {location.href}, location.origin:
        {location.origin}
    </>
  );
}

export default withLocation(Location);
```

In the consumer of `Location`, what we import as `Location` is the default export – that is, `withLocation(Location)`:

```
import Location from './location';

function App() {
  return <Location>;
}
```

The `Location` component renders `location.href` and `location.origin`, based on what's provided by `withLocation`.

location.href: http://localhost:1234/, location.origin:
http://localhost:1234

Figure 4.15: The Location component rendering the href and origin, based on what's provided by withlocation

We've now seen a simple example of a key benefit of higher-order components, which is that the component doing the rendering doesn't need to be directly aware of where to receive information; it can read props instead.

Use cases

The `withLocation` example already showed a simple reason why we could use higher-order components – to maintain separation of concerns.

In our Location component example, it's completely possible for Location to access `window.location` directly. What that would mean, however, is that the Location component is aware of global objects, which could be undesirable. For example, it might make unit-testing of `Location` more difficult, since it's accessing something beyond its props.

Limitations

As with all abstractions, higher-order components are a layer of indirection. This means that tracing where a prop comes from can be more difficult than when props are passed explicitly from the parent component.

Tracing the props becomes even more difficult when the higher-order component comes from a third-party library (and thus is harder to inspect).

Higher-order components can have a cost in terms of rendering in the browser, since we wrap our component in another component if we stack too many higher-order components on top of each other.

For example, the following `ConnectedComponent` uses three higher-level components:

```
const ConnectedComponent = withRouter(
  withHttpClient(withAnotherDependency
    (ComponentWithDependencies))
);
```

As a consumer of `ConnectedComponent`, we'll likely render four components – the ones provided by `withRouter`, `withHttpClient`, `withAnotherDependency`, and `ComponentWithDependencies`. If we had another way to inject the router, HTTP client and another dependency, we could reduce the number of components to one, only needing `ComponentWithDependencies`.

This drawback leads us to the next topic in this chapter – hooks. Hooks provide us a way to access data and logic in similar scenarios as higher-order components, without additional components being rendered. Hooks are a great replacement for logic-heavy higher-order components.

The hooks pattern

We've now covered what might be considered *legacy* patterns in React – render props and higher-order components.

> *You'll note that the React documentation page about higher-order components has the following disclaimer: "Higher-order components are not commonly used in modern React code."*

> **Additional reading**
>
> The React documentation for `useState` and `useEffect` hooks:
>
> `useState:` https://react.dev/reference/react/useState
>
> `useEffect:` https://react.dev/reference/react/useEffect

So, what we know so far is that hooks allow us to do what we did with render props and that higher-order components are not recommended any more. This is because hooks provide a way to access all the React primitives, including state and the component life cycle.

React provides built-in hooks. The two we'll focus on are `useState` and `useEffect`. One key feature of hooks is that we can write custom hooks that build on top of React built-in hooks and other custom hooks, which means we have a new way to share logic in React.

An implementation/example

We'll implement simple data fetching using the class React components, and then hooks. This will showcase how state and life cycle events are handled in both cases.

We'll start with the class components. The regular way to implement data fetching is by using life cycle hooks; the initial one tends to be componentDidMount.

Our BasketItemsClassical component takes httpClient and basketId.

The component's constructor initializes a state.basketSession variable to an empty object, { }:

```
import React from 'react';

export class BasketItemsClassical extends React.Component {
  constructor(props) {
    super(props);
    this.state = {
      basketSession: {},
    };
  }
}
```

Next, we'll add a setBasketSession method, which will call this.setState to set basketSession as the passed parameter.

We'll also add componentDidMount, which calls httpClient.get() with the fakestoreapi. com URL to load carts, using the basketId prop:

```
export class BasketItemsClassical extends React.Component {
  // no change to the constructor
  componentDidMount() {
    this.props.httpClient
      .get(`https://fakestoreapi.com/carts/${this.props.basketId}`)
      .then((session) => this.setBasketSession(session));
  }
  setBasketSession(session) {
    this.setState({ basketSession: session });
  }
}
```

What this now means is that we should be able to render out the contents of this.state. basketSession in the component's render() method:

```
export class BasketItemsClassical extends React.Component {
  // no change to the constructor, componentDidMount or
    setBasketSession
```

```
  render() {
    return <pre>{JSON.stringify(this.state.basketSession,
      null, 2)}</pre>;
  }
}
```

Our `BasketItemsClassical` can be used as follows by passing `httpClient` and `basketId` as props:

```
export function BasketClassical({ basketId, httpClient }) {
  return (
    <form>
      <fieldset>
        <label>Class</label>
        <BasketItemsClassical basketId={basketId}
            httpClient={httpClient} />
      </fieldset>
    </form>
  );
}
```

`BasketClassical` can then be used in our App, as follows:

```
const httpClient = {
  async get(url) {
    const response = await fetch(url);
    return await response.json();
  },
};

function App() {
  return (
    <>
      <BasketClassical basketId="5" httpClient={httpClient} />
    </>
  );
}
```

In the browser, this displays as follows:

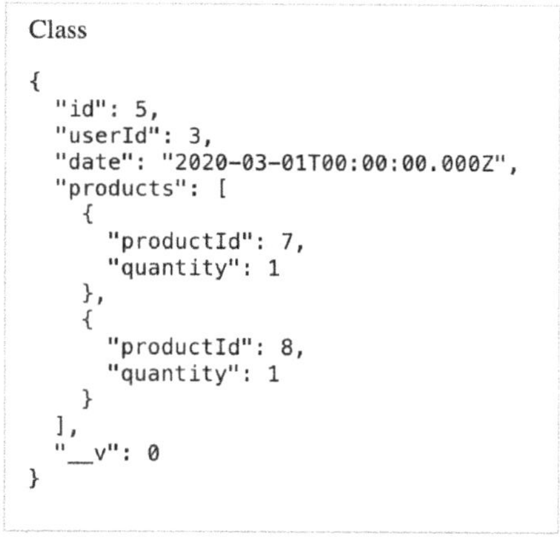

```
Class

{
  "id": 5,
  "userId": 3,
  "date": "2020-03-01T00:00:00.000Z",
  "products": [
    {
      "productId": 7,
      "quantity": 1
    },
    {
      "productId": 8,
      "quantity": 1
    }
  ],
  "__v": 0
}
```

Figure 4.16: The Basket class component loading JSON data

Here's the same example with hooks; instead of using `componentDidMount`, we can use the `useEffect` hook, and instead of `this.state` in the constructor and `this.setState`, we use the `useState` hook. In order to use hooks, we use a React function component (React class components don't support hooks):

```
export function BasketItemsHooks({ basketId, httpClient }) {
  const [basketSession, setBasketSession] = useState({});
  useEffect(() => {
    httpClient
      .get(`https://fakestoreapi.com/carts/${basketId}`)
      .then((session) => setBasketSession(session));
  }, []);
  return <pre>{JSON.stringify(basketSession, null, 2)}</pre>;
}
```

Our `BasketItemsHooks` can be used in the same way as `BasketItemsClassical`, by passing `httpClient` and `basketId` as props:

```
export function BasketHooks({ basketId, httpClient }) {
  return (
    <form>
      <fieldset>
        <label>Hooks</label>
        <BasketItemsHooks basketId={basketId}
          httpClient={httpClient} />
      </fieldset>
```

```
    </form>
  );
}
```

We'll also need to modify `App` to render `BasketHooks` in addition to `BasketClassical`:

```
// no change to httpClient
function App() {
  return (
    <>
      <BasketClassical basketId="5" httpClient={httpClient} />
      <BasketHooks basketId="5" httpClient={httpClient} />
    </>
  );
}
```

Both `BasketHooks` (*Figure 4.17*) and `BasketClassical` (*Figure 4.16*) yield the same JSON output after the HTTP requests are completed.

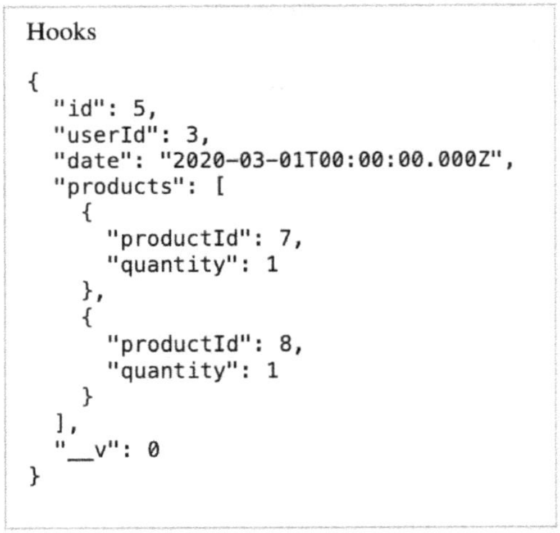

```
Hooks

{
  "id": 5,
  "userId": 3,
  "date": "2020-03-01T00:00:00.000Z",
  "products": [
    {
      "productId": 7,
      "quantity": 1
    },
    {
      "productId": 8,
      "quantity": 1
    }
  ],
  "__v": 0
}
```

Figure 4.17: The Hooks basket loading JSON data

The hooks approach is slightly more compact; each part of the functionality does feel a bit more self-contained. For example, the initial state is handled in the same place that defines what the state update function will be in the hooks version. In the class example, the `initialisation` state is in the constructor, and the state update function is a method. In the `BasketClassical` example, there was the option to simplify the component by removing the state update method and using a direct `this.setState({ bookingSession: session })` call.

Use cases

A simple way to think about hooks and class or function components is as follows:

- Hooks for shared logic
- Components for logic that is related to rendering

The higher-order component and render prop patterns, which are used to separate presentation and business logic, are unlikely to be needed any more and can be replaced by custom hooks.

React hooks and function components are the recommended way to develop modern React applications.

Limitations

Hooks can't be used in React class components, as detailed in the React documentation: `https://react.dev/reference/react/Component#defining-a-class-component`. Note that function components are the recommended way to build React components.

In code bases with heavy usage of class components, higher-order components should remain in use instead of migrating components to functions in order to use hooks.

The last piece of the puzzle with React components is how to bypass the prop drilling problem and pass data without changing each component in a React component tree. The pattern we use for this is the provider pattern, which we'll cover in the next section.

The provider pattern

The provider pattern in React is where one component in the tree makes data accessible to all its descendants. This is usually accomplished using the React **Context** primitive.

Use case – the prop drilling problem

The key use case for the provider pattern is to avoid the **prop drilling** problem.

A large majority of the time, a component's main input is the prop it receives from its parent component. A state management pattern to share state between components in React is to *lift state up*. Lifting state up means to store relevant state in a common ancestor of the components that require the shared state.

As stated in the React.js docs (`https://react.dev/learn/sharing-state-between-components`)

> *When you want to coordinate two components, move their state to their common parent. Then pass the information down through props from their common parent*

This can lead to **prop drilling** when the common parent has multiple components between it and the components requiring the props. This means all the intermediate components will receive the props, but they will only use them to forward them on to the next layer of components.

As stated in the React.js docs (`https://react.dev/learn/passing-data-deeply-with-context#the-problem-with-passing-props`)

> *Passing props is a great way to explicitly pipe data through your UI tree to the components that use it. But passing props can become verbose and inconvenient when you need to pass some prop deeply through the tree, or if many components need the same prop. The nearest common ancestor could be far removed from the components that need data, and lifting state up that high can lead to a situation called "prop drilling*

The provider pattern is a solution to the prop drilling problem, since every descendent of the provider component will have access to the data it provides.

An implementation/example

Let's look back to the examples from *The hooks pattern* section, where we had the App rendering `BasketClassical` and `BasketHooks`, which render `BasketItemsClassical` and `BasketItemsHooks`, respectively.

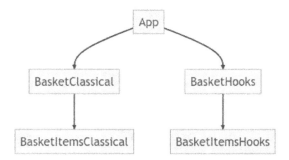

Figure 4.18: A React app tree with BasketClassical, BasketHooks, and their descendants

This illustrates the prop drilling problem, since `BasketClassical` and `BasketHooks` don't use `basketId` or `httpClient` beyond passing it to `BasketItemsClassical` and `BasketItemsHooks`.

There are multiple ways to consume a context in React, but it all begins by creating a context:

```
import React, { createContext } from 'react';

const HttpClientContext = createContext(null);
```

```
export function HttpClientProvider({ httpClient, children
  }) {
  return (
    <HttpClientContext.Provider value={httpClient}>
      {children}
    </HttpClientContext.Provider>
  );
}
```

`HttpClientContext` is a context that's initialized with the null value. `HttpClientProvider` is a component that takes a `httpClient` value, setting it as the value that `HttpClientContext.Provider` will pass to its descendants in the component tree.

In order to use `HttpClientContext`, we can use `HttpClientContext.Consumer`:

```
export const HttpClientConsumer = HttpClientContext.Consumer;
```

`HttpClientContext.Consumer` has a children render-prop (function) that takes the value of the context (in this case, `httpClient`) and returns some JSX to render:

```
// no change to httpClient
function App() {
  return (
    <HttpClientProvider httpClient={httpClient}>
      {/* what's below could be however deep in the
          component tree */}
      <HttpClientConsumer>
        {(httpClient) => (
          <BasketItemsClassical basketId="5" httpClient=
          {httpClient} />
        )}
      </HttpClientConsumer>
    </HttpClientProvider>
  );
}
```

This yields the following output in the browser:

```
{
  "id": 5,
  "userId": 3,
  "date": "2020-03-01T00:00:00.000Z",
  "products": [
    {
      "productId": 7,
      "quantity": 1
    },
    {
      "productId": 8,
      "quantity": 1
    }
  ],
  "__v": 0
}
```

Figure 4.19: The JSON contents of basketId=5 from fakestoreapi.com

The approach using `HttpClientContext.Consumer` directly is a bit unwieldy. Instead, we could wrap it in a higher-order component, `withHttpClient`, which consumes `HttpClientConsumer`. The benefit here is that we only have one place that uses `HttpClientConsumer`:

```
export function withHttpClient(Component) {
  return (props) => (
    <HttpClientConsumer>
      {(httpClient) => <Component {...props} httpClient=
        {httpClient} />}
    </HttpClientConsumer>
  );
}
```

With a slight difference from our example of a higher-order component, we'll export const `Connected BasketItemsClassical` with the value `withHttpClient(BasketItemsClassical)`. The *connected* nomenclature is a call back to the large React Redux code bases where the components are often split among *presentation* and *container* components. The Redux higher-order component is called `connect`, and all the containers are connected:

```
export const ConnectedBasketItemsClassical =
  withHttpClient(BasketItemsClassical);
```

We can then use `ConnectedBasketItemsClassical` as follows. Note that we don't pass an `httpClient` prop:

```
function App() {
  return (
    <HttpClientProvider httpClient={httpClient}>
      {/* what's below could be however deep in the
          component tree */}
      <ConnectedBasketItemsClassical basketId="5" />
    </HttpClientProvider>
  );
}
```

The higher-order component version using `withHttpClient` outputs the same value as the direct `HttpClientConsumer` implementation.

```
{
  "id": 5,
  "userId": 3,
  "date": "2020-03-01T00:00:00.000Z",
  "products": [
    {
      "productId": 7,
      "quantity": 1
    },
    {
      "productId": 8,
      "quantity": 1
    }
  ],
  "__v": 0
}
```

Figure 4.20: The JSON contents of basketId=5 from fakestoreapi.com

The final approach to using context and the provider pattern is to leverage the React `useContext` hook. Similar to how `HttpClientContext.Consumer` allows us to access the context provider's value, the hook fulfills that same role. So, the output of `useContext(context)` is the current value, based on where the hook renders in the component tree.

It's customary to wrap the `useContext` hook in a more descriptive name (as we did for `HttpClientContext.Consumer`):

```
import React, { createContext, useContext } from 'react';

// no changes to HttpClientContext definition or
```

```
    HttpClientContextConsumer

export function useHttpClient() {
  const httpClient = useContext(HttpClientContext);
  return httpClient;
}
```

This time, using `httpClient` from the `HttpClientContext` requires component-level changes. So, we'll write the following implementation of `BasketItemsHooksUseContext`:

```
export function BasketItemsHooksUseContext({ basketId }) {
  const httpClient = useHttpClient();
  const [basketSession, setBasketSession] = useState({});
  useEffect(() => {
    // @ts-ignore
    httpClient
      .get(`https://fakestoreapi.com/carts/${basketId}`)
      .then((session) => setBasketSession(session));
  }, []);
  return <pre>{JSON.stringify(basketSession,
   null, 2)}</pre>;
}
```

`BasketItemsHooksUseContext` can be used as follows. Note that we're not passing `BasketItemsHooksUseContext`, a `httpClient` prop:

```
function App() {
  return (
    <HttpClientProvider httpClient={httpClient}>
      {/* what's below could be however deep in the
      component tree */}
      <BasketItemsHooksUseContext basketId="5" />
    </HttpClientProvider>
  );
}
```

This implementation is yet again equivalent to the previous implementations we did with `HttpClientConsumer` and `HttpClient`.

```json
{
  "id": 5,
  "userId": 3,
  "date": "2020-03-01T00:00:00.000Z",
  "products": [
    {
      "productId": 7,
      "quantity": 1
    },
    {
      "productId": 8,
      "quantity": 1
    }
  ],
  "__v": 0
}
```

Figure 4.21: The JSON contents of basketId=5 from fakestoreapi.com

We saw how to use the provider pattern to solve the prop drilling problem in React apps. Let's now look at some limitations of this pattern in the next section.

Limitations

The provider pattern is a layer of indirection. It might not always be obvious where a context's value is coming from, or it might sometimes be necessary to change the provider/context shape to make some changes at the component level. For example, when using context with hooks, the hook shows a direct link between consuming component and the context, but it doesn't necessarily show the provider or how the value inside the context is defined.

It's also sometimes possible to solve the prop drilling problem by making liberal use of children and composing the components in a single large JSX return, such as the following:

```jsx
function MyComponent() {
  return <ContainerComponent requiredProp={'value'}>
    <OtherComponent prop="other-value"/>
    <FinalComponent prop="final-value"/>
  </ContainerComponent>
}
```

In MyComponent, we pass the props directly to OtherComponent and FinalComponent from MyComponent. If we had ContainerComponent encapsulating OtherComponent and FinalComponent, the props would be drilled via ContainerComponent (it doesn't use the props but receives them, in order to pass them to its descendants).

Summary

In this chapter, we looked at how reactive view library patterns enable us to build React applications more effectively when the component paradigm starts to break down.

The render prop pattern allows us to decouple data logic and rendering logic by yielding rendering control back to the consumer of a component.

The higher-order component pattern allows components to implement logic (data or rendering) against their props, without having to concern themselves with where the information comes from.

The hooks pattern means that React primitives that were only available in class components are now available as self-contained logic chunks to function components. Hooks can be composed separately of components, which makes hooks a powerful primitive and can partly replace the render prop and higher-order component patterns.

The provider pattern allows React components to pass data not only to their children but also to any descendent component.

Now that we're familiar with reactive view library patterns, in the next chapter, we will look at rendering and page hydration strategies to improve web applications' performance.

5

Rendering Strategies and Page Hydration

Rendering strategies and page hydration approaches allow us to leverage the JavaScript client and server ecosystem to serve performant and scalable web applications, depending on the needs of our end users. The React and JavaScript techniques covered in this chapter are another set of tools to augment the *Chapter 4* chapter. We'll use the strengths of the client (browser) and server (specifically, Node.js) runtimes to deliver fast and scalable React websites to users.

In this chapter, we'll cover the following topics:

- What the trade-offs are between client and server rendering of React applications by implementing pure client and server rendering applications
- The types of advantages that frameworks such as Next.js can bring with static site generation functionality, alongside server-side rendering
- Bridging the client-server rendering gap with a React page rehydration example and its gotchas
- Streaming server rendering in React

By the end of this chapter, you'll be able to select appropriate rendering and page hydration strategies with React and be able to implement framework-level functionality, enabling you to make better technology choices.

Technical requirements

You can find the code files for this chapter on GitHub at `https://github.com/PacktPublishing/Javascript-Design-Patterns`

Client and server rendering with React

In a web context, client-side rendering is the process by which JavaScript is used inside a user's browser to generate or update the page contents. A fully client-side-rendered application will only display meaningful content when the relevant JavaScript code has completed downloading, parsing, and running.

In the following sequence diagram, we use the term "origin" instead of something such as "server," since one benefit of full client-side rendering is that the resources "serving" our content can be what's called *static hosting*. This includes services such as **AWS Simple Storage Service (S3)**, Netlify, Cloudflare Pages, and GitHub Pages, among others. There's no dynamic server-side component in these services.

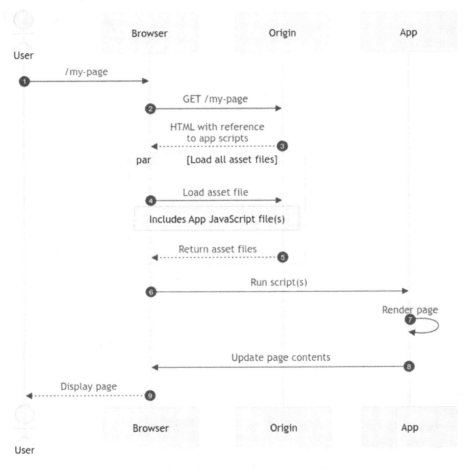

Figure 5.1: A client-side-rendering sequence diagram

In contrast, server-side rendering denotes the process by which a server generates a full HTML document when a browser requests it and returns it.

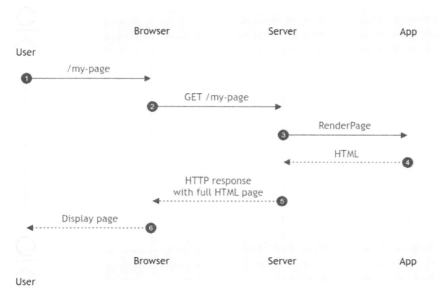

Figure 5.2: A server-side-rendering sequence diagram

Client-side rendering in React

Client rendering is the default rendering method with React. Let's build an application that renders client-side from scratch:

1. We start with an App component that renders some text and its `type` prop:

    ```
    export function App({ type = '' }) {
      return (
        <div>
          <p>Hello from the {type + ' '}app</p>
        </div>
      );
    }
    ```

2. We then create an entry point file, `client.jsx`, which imports the app and uses `ReactDOM` to render it, with the `type` prop set to `"client render"`.

    ```
    import React from 'react';
    import ReactDOM from 'react-dom/client';
    import { App } from './src/app';
    ```

```
ReactDOM.createRoot(document.querySelector
  ('#app')).render(
  <App type={`"client render"`} />
);
```

3. In order for this example to run, we need an HTML document that allows `ReactDOM.createRoot` to run successfully. In other words, we need an HTML document that has an element with `id=app` and references our entry point:

```
<div id="app"></div>
<script src="./dist/client.js"></script>
```

4. Note that the entry point is `dist/client.js` instead of `client.jsx`. That's due to the fact that React's JSX syntax can't be run natively in the browser. Instead, we run our entry point file, `client.jsx`, through a compilation and bundling step using `esbuild`. Our build command looks something like this:

```
npx esbuild client.jsx --bundle --outdir=dist
```

Now, if we load the `index.html` file in the browser, we see the following:

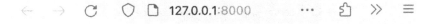

Hello from the "client render" app

Figure 5.3: Hello from the "client render" app rendered in the browser

Server rendering in React

Node.js, introduced on its website as "an open-source, cross-platform JavaScript runtime environment," gives us the ability to run JavaScript on a server. A commonly used package to build servers in Node.js is Express.

In this section, we'll see how to use Node.js and Express to server-render a React application.

A simple Express server that returns `'Server-rendered hello'` when the root path is loaded looks as follows:

```
import express from 'express';
const app = express();
app.get('/', (_req, res) => {
  res.send('Server-rendered hello');
});
```

```
const { PORT = 3000 } = process.env;
app.listen(PORT, () => {
  console.log(`Server started on
    http://localhost:${PORT}`);
});
```

Again, we'll use `esbuild` to bundle and compile the JSX to JavaScript:

```
npx esbuild server.js --bundle --platform=node --outdir=dist
```

We can then start the server using the following:

```
node dist/server.js
```

By default, it runs on port 3000, but that can be overridden with environment variables.

When we load `localhost:3000`, we see this message in the browser.

Server-rendered hello

Figure 5.4: Server-rendered hello rendered in the browser

This is a really minimal example of server-rendering using Node.js and Express.

Next, we'll see how to leverage the `ReactDOM` package to take React components and server-render them:

1. The `ReactDOM` package provides both a `react-dom/client` entry point (which we used in the previous section) and a `react-dom/server` entry point. As the names allude to, the client entry point is meant to be used on the client (in the browser, "client-side" JavaScript), and the server entry point is meant to be used on the server (via Node.js or another server-side JavaScript runtime).

2. The two methods we'll use are `ReactDOMServer.renderToStaticMarkup` and `ReactDOMServer.renderToString`; these two methods will allow us to server-render our React application to HTML.

3. In a `src/server-render.jsx` file, we have the following `renderNav` and `serverRenderApp` functions, which use `ReactDOMServer.renderToStaticMarkup` and `ReactDOMServer.renderToString` to render Nav and App, respectively:

    ```
    import React from 'react';
    import ReactDOMServer from 'react-dom/server';
    import { App } from './app';
    ```

```
import { Nav } from './nav';

export function renderNav() {
  return ReactDOMServer.renderToStaticMarkup(<Nav />);
}
export function serverRenderApp() {
  return ReactDOMServer.renderToString(<App
    type={`"server render"`} />);
}
```

Here, app.jsx is the same as we had in the previous section, and nav.jsx is as follows:

```
import React from 'react';

export function Nav() {
  return (
    <ul>
      <li>
        <a href="/">Server-render only</a>
      </li>
    </ul>
  );
}
```

4. We can then use renderNav and serverRenderApp in server.js. We modify the app.get('/') handler to render the navigation and the app:

```
// no other changes
app.get('/', (_req, res) => {
  res.send(`
    <!DOCTYPE html>
    ${renderNav()}
    <h1>Server-render only</h1>
    <div id="app">${serverRenderApp()}</div>
  `);
});
```

5. When we rebuild the server, we run `node dist/server.js` and open `localhost:3000` to see the following:

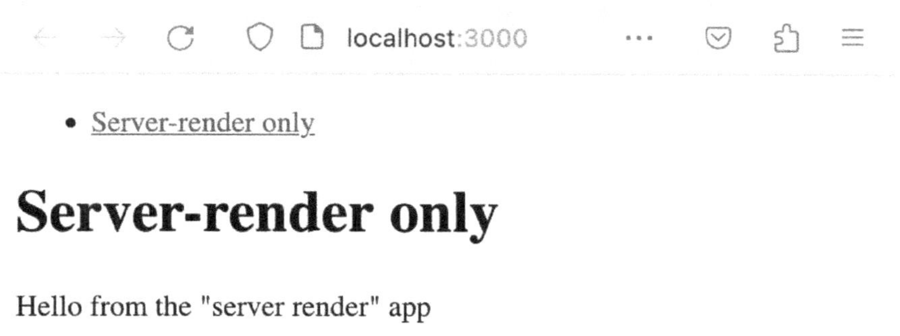

*   Server-render only

Server-render only

Hello from the "server render" app

Figure 5.5: Hello from the "server render" app being rendered alongside a heading and the nav

What's the difference between `ReactDOMServer.renderToStaticMarkup` and `ReactDOMServer.renderToString`? The short answer is that `renderToStaticMarkup` can't be rehydrated client-side; in other words, it can't be used as the initial HTML, and then the same React application code can be run client-side to give a fully interactive experience. We'll revisit this in a later section of the chapter.

Trade-offs between client and server rendering

So, what are the benefits and drawbacks of client and server rendering?

Client rendering's main benefit is that the application's "work" is done fully in the user's browsers, which makes it highly scalable since the amount of users using the system will not put pressure on the origin servers. Client rendering's main drawbacks relate to functionality that is only available on the server side – for example, server-side-only cookies or setting the `meta` tags for social media preview.

Server rendering's main drawback is that work has to happen on the server. As stated previously, the server being a "controlled" environment has some benefits, namely its latency to other co-located systems will tend to be lower than a full browser-server round trip, since the server's network is known and unlikely to have as much variance in performance as an end user's network. By not waiting for a full-page load, followed by an asset load, followed by a JavaScript "parse and execute" cascade, server-rendering can improve "core web vitals", such as **largest contentful paint (LCP)** and **cumulative layout shift (CLS)**.

Ultimately, client-rendered functionality is a key reason we use JavaScript, which means removing that ability will only make sense in constrained use cases such as content sites (e.g., blogs, news sites, and documentation sites).

We've now seen the difference between client and server rendering, as well as how to implement both with React and Node.js. In the following section, we'll look at rendering approaches enabled by the Next.js framework for React.

Static rendering with Next.js

Next.js is a React framework for creating full stack web applications. What this means is that it provides tools and opinions that will help developers be more productive in the short and long term.

Next.js includes a filesystem router for "pages", a set of routing primitives for React, support for client and server rendering, and data fetching primitives, among others.

The features of Next.js we'll focus on are the **static site generation** (SSG) ones. This type of rendering methodology resembles server rendering but mitigates some of its drawbacks, since the rendering pass is done at build time instead of at request time.

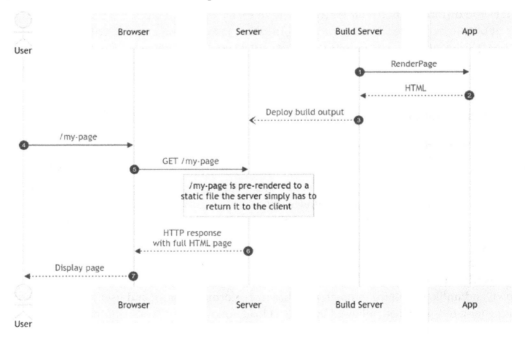

Figure 5.6: A sequence diagram for a pre-rendered/static site generation use case

Now that we've looked at how static site generation changes the data flow when a user requests a website, we'll look at Next.js automatic static generation.

Automatic static generation

In Next.js, the filesystem-based routing means that each path in your web application corresponds to a file in the `pages` directory of your application. For example, `/` corresponds to `pages/index.js`.

Next.js defaults to static generation when no Next.js data fetching methods are used for a given page. You can find more information from the Next.js documentation – *Automatic Static Optimization* (`https://nextjs.org/docs/pages/building-your-application/rendering/automatic-static-optimization`).

> *Next.js automatically determines that a page is static (i.e., can be prerendered) if it has no blocking data requirements. This determination is made by the absence of getServerSideProps and getInitialProps on the page.*

For example, the following page in a Next.js application will be statically generated, since it only exports a page component (the default export of `Index`); no `getServerSideProps` or `getInitialProps` function is exported:

```
import React from 'react';
import Head from 'next/head';
import Link from 'next/link';

export default function Index() {
  return (
    <>
      <Head>
        <title>Next Static Rendering - Automatic Static
          Generation</title>
        <meta name="viewport" content="width=device-width,
          initial-scale=1" />
      </Head>
      <main>
        <ul>
          <li>
            <Link href="/products">Products Page (SSG)
            </Link>
          </li>
          <li>
            <Link href="/cart">Cart Page (SSR)</Link>
          </li>
        </ul>
      </main>
    </>
  );
}
```

We can see this during next build in the following screenshot; / route (page) is marked as Static in the output:

```
npx next build

info  - Linting and checking validity of types
info  - Creating an optimized production build
info  - Compiled successfully
info  - Collecting page data
info  - Generating static pages (3/3)
info  - Finalizing page optimization

Route (pages)                              Size        First Load JS
┌ ○ /                                      2.73 kB        75.8 kB
└ ○ /404                                    182 B         73.2 kB
+ First Load JS shared by all              73.1 kB
  ├ chunks/framework-fcfa81c6fe8caa42.js   45.2 kB
  ├ chunks/main-7039e34bfb6f1a68.js        26.9 kB
  ├ chunks/pages/_app-c7a111f3ee9d686c.js  195 B
  └ chunks/webpack-8fa1640cc84ba8fe.js     750 B

○  (Static)   automatically rendered as static HTML (uses no initial
props)
```

When we run the built Next.js output with next start, the page behaves as expected.

- Products Page (SSG)
- Cart Page (SSR)

Figure 5.7: Links to the Products and Cart pages rendering

This example is a relatively constrained use case since we have no dynamic data fetching requirements. It still showcases Next.js defaulting to static rendering if the page does not use any functionality that excludes static generation. For more advanced use cases, Next.js also allows use of "build-time" dynamic data, which means we can use a third-party data source to generate the page content, and more.

We've seen how Next.js defaults to automatic static generation. Next, we'll see how to configure a Next.js page to load data to render a page as static.

Static generation with a third-party data source

Next.js has a `getStaticProps` data fetching method that allows us to load data at build time, which will be passed to a page.

The following sequence diagram illustrate what this involves:

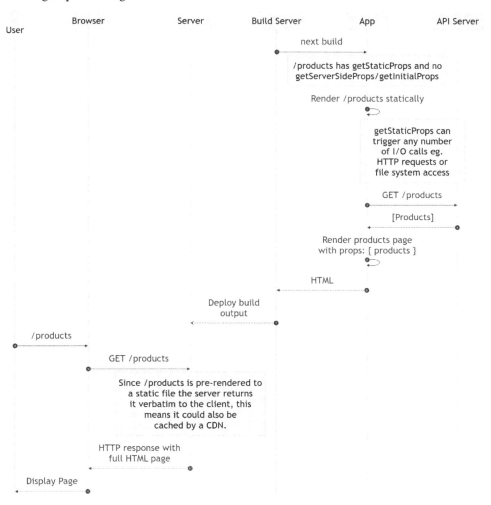

Figure 5.8: A sequence diagram of Next.js pre-rendering using getStaticProps

For example, if we want to build a "product list" page based on fakestoreapi.com data, we can write the following getStaticProps method in a pages/products/index.js page:

```
export async function getStaticProps() {
  const products = await fetch
    ('https://fakestoreapi.com/products').then(
    (res) => res.json()
  );
  return {
    props: {
      products,
    },
  };
}
```

Here's a product example in the response to illustrate the data shape:

```
{
  id: 1,
  title: 'Fjallraven - Foldsack No. 1 Backpack, Fits 15
    Laptops',
  price: 109.95,
  description: 'Your perfect pack for everyday use and
  walks in the forest. Stash your laptop (up to 15 inches)
  in the padded sleeve, your everyday',
  category: "men's clothing",
  image: 'https://fakestoreapi.com/img/
    81fPKd-2AYL._AC_SL1500_.jpg',
  rating: { rate: 3.9, count: 120 }
}
```

Based on the data provided by getStaticProps, we can build a ProductIndexPage component. We'll loop through each product from props.products and render them in an unordered list. Each item will include a link to the /products/[id] page (which doesn't exist yet):

```
import React from 'react';
import Link from 'next/link';
import Head from 'next/head';

export default function ProductIndexPage({ products }) {
  return (
    <>
      <Head>
        <title>Products</title>
      </Head>
```

```
    <div>
      <h2>Products</h2>
      <ul>
        {products.map((product) => {
          return (
            <li key={product.id}>
              <Link
                href={{
                  pathname: '/products/[id]',
                  query: { id: product.id },
                }}
              >
                {product.title}
              </Link>
            </li>
          );
        })}
      </ul>
    </div>
  </>
  );
}
// no change to getStaticProps
```

This page will now be built when `next build` is run. As we can see from the output, the `/products` page is marked as SSG (static site generation):

```
info  - Linting and checking validity of types...
info  - Creating an optimized production build...
info  - Compiled successfully
info  - Collecting page data
info  - Generating static pages (4/4)
info  - Finalizing page optimization

Route (pages)                              Size      First Load JS
┌ ○ /                                      464 B         75.9 kB
├ ○ /404                                   182 B         73.2 kB
└ ● /products                              426 B         75.9 kB
+ First Load JS shared by all              73.1 kB
  ├ chunks/framework-fcfa81c6fe8caa42.js   45.2 kB
  ├ chunks/main-7039e34bfb6f1a68.js        26.9 kB
  ├ chunks/pages/_app-c7a111f3ee9d686c.js  195 B
  └ chunks/webpack-8fa1640cc84ba8fe.js     750 B
```

```
○  (Static)    automatically rendered as static HTML (uses no initial
props)
●  (SSG)       automatically generated as static HTML + JSON (uses
getStaticProps)
```

When we start the Next.js server with `next start` and navigate to `/products`, we see the following. Note that the products on the page won't change unless we rebuild the application.

Products

- Fjallraven - Foldsack No. 1 Backpack, Fits 15 Laptops
- Mens Casual Premium Slim Fit T-Shirts
- Mens Cotton Jacket
- Mens Casual Slim Fit
- John Hardy Women's Legends Naga Gold & Silver Dragon Station Chain Bracelet
- Solid Gold Petite Micropave
- White Gold Plated Princess
- Pierced Owl Rose Gold Plated Stainless Steel Double
- WD 2TB Elements Portable External Hard Drive - USB 3.0
- SanDisk SSD PLUS 1TB Internal SSD - SATA III 6 Gb/s
- Silicon Power 256GB SSD 3D NAND A55 SLC Cache Performance Boost SATA III 2.5
- WD 4TB Gaming Drive Works with Playstation 4 Portable External Hard Drive
- Acer SB220Q bi 21.5 inches Full HD (1920 x 1080) IPS Ultra-Thin
- Samsung 49-Inch CHG90 144Hz Curved Gaming Monitor (LC49HG90DMNXZA) – Super Ultrawide Screen QLED
- BIYLACLESEN Women's 3-in-1 Snowboard Jacket Winter Coats
- Lock and Love Women's Removable Hooded Faux Leather Moto Biker Jacket
- Rain Jacket Women Windbreaker Striped Climbing Raincoats
- MBJ Women's Solid Short Sleeve Boat Neck V
- Opna Women's Short Sleeve Moisture
- DANVOUY Womens T Shirt Casual Cotton Short

Figure 5.9: The Products list page is statically pre-rendered with products from fakestoreapi.com

We've seen how to use `getStaticProps` to generate pages based on a third-party API, but how would we generate the `/products/[id]` pages ahead of them being requested? To do that, we need to be able to provide the "required paths" (or URLs) that Next.js needs to generate. This is what we'll look at in the following section.

Static generation with dynamic paths

It can be useful to pre-generate pages with dynamic paths and contents.

We could use getServerSideProps and render the pages on demand. In the context that we're working in, that would be valid for a "cart" page.

getServerSideProps is server-side rendering, as we've seen previously. The reason a cart page should probably be server-rendered is that it can change very quickly, based on end user interaction. An example of a page that is dynamic but wouldn't change quickly based on an end user action is a "view single product" page. We'll see how to statically generate that after the cart page example.

We create a pages/cart.js file, where we provide the following getServerSideProps, which loads the cart, figures out the relevant product IDs (per cart content), and loads them (in order to display some information about them):

```
export async function getServerSideProps({ query }) {
  const { cartId = 1 } = query;
  const cart = await fetch(`https://fakestoreapi.com/
carts/${cartId}`).then(
    (res) => res.json()
  );
  const productsById = (
    await Promise.all(
      cart.products.map(async (product) => {
        return await fetch(
          `https://fakestoreapi.com/products/
            ${product.productId}`
        ).then((res) => res.json());
      })
    )
  ).reduce((acc, curr) => {
    acc[curr.id] = curr;
    return acc;
  }, {});
  return {
    props: {
      cart,
      productsById,
    },
  };
}
```

We can then build a page component and make it the default export. In the component, we loop through the cart products, rendering some count information and some product information, based on `props.productsById`:

```
import Head from 'next/head';
import React from 'react';
export default function CartPage({ cart, productsById }) {
  return (
    <>
      <Head>
        <title>Cart Page</title>
      </Head>
      <div>
        <ul>
          {cart.products.map((product) => {
            return (
              <li key={product.productId}>
                {product.quantity} x {productsById
                  [product.productId]?.title}
              </li>
            );
          })}
        </ul>
      </div>
    </>
  );
}
```

We know this is a server-side rendered page because when we run `next build`, it gets marked as such (and doesn't increase the `Generating static pages` count):

```
npx next build

info  - Linting and checking validity of types
info  - Creating an optimized production build
info  - Compiled successfully
info  - Collecting page data
info  - Generating static pages (4/4)
info  - Finalizing page optimization

Route (pages)                           Size      First Load JS
┌ ○ /                                   464 B          75.9 kB
├ ○ /404                                182 B          73.2 kB
├ λ /cart                               445 B          73.5 kB
```

```
L • /products                                426 B            75.9 kB
+ First Load JS shared by all                73.1 kB
  ├ chunks/framework-fcfa81c6fe8caa42.js      45.2 kB
  ├ chunks/main-7039e34bfb6f1a68.js           26.9 kB
  ├ chunks/pages/_app-c7a111f3ee9d686c.js     195 B
  L chunks/webpack-8fa1640cc84ba8fe.js        750 B

λ  (Server)   server-side renders at runtime (uses
   getInitialProps or getServerSideProps)
○  (Static)   automatically rendered as static HTML (uses no
   initial props)
•  (SSG)       automatically generated as static HTML + JSON
   (uses getStaticProps)
```

We can load the /carts page with a ?cartId=1 query param and see Cart 1.

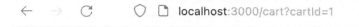

- 4 x Fjallraven - Foldsack No. 1 Backpack, Fits 15 Laptops
- 1 x Mens Casual Premium Slim Fit T-Shirts
- 6 x Mens Cotton Jacket

Figure 5.10: The cart page with Cart 1 loaded and contents displaying

We can also load the /carts page with cartId=3 query param and see Cart 3.

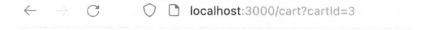

- 2 x Fjallraven - Foldsack No. 1 Backpack, Fits 15 Laptops
- 1 x WD 2TB Elements Portable External Hard Drive - USB 3.0

Figure 5.11: Cart page with Cart 3 loaded and contents displaying

We've now seen how to render the cart page on demand; a page that we mentioned is a good fit for build-time pre-rendering (i.e., static site generation) is the products/[id] page. In order to render this page, we need to provide the "paths" that Next.js needs to attempt to pre-render, since [id] is dynamic.

The following diagram shows how getStaticPaths and getStaticProps interact with each other. In short, getStaticPaths returns a list of "paths"; getStaticProps is then called on each item in that list of paths and can make the relevant I/O calls to provide the page's props.

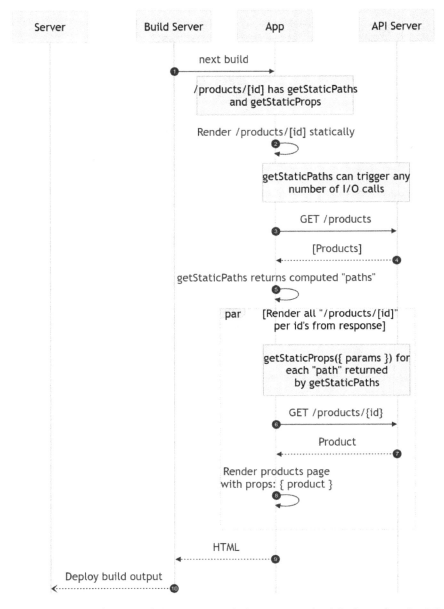

Figure 5.12: Sequence diagram of Next.js pre-rendering using getStaticPaths and getStaticProps

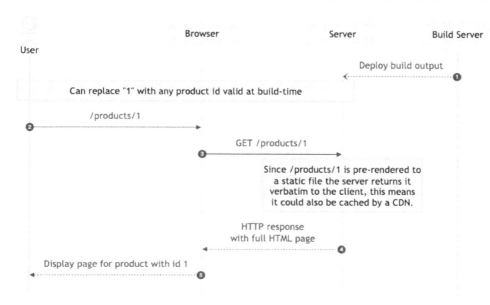

Figure 5.13: Sequence of a request to a pre-rendered Next.js app

In our sample Next.js app, we can create a `pages/products/[id].js` file with the following `getStaticPaths` and `getStaticProps` functions:

```
export async function getStaticPaths() {
  const products = await fetch('https://fakestoreapi.com/
    products')
    .then((res) => res.json())
    .then((json) => json);

  const paths = products.map((product) => ({
    params: { id: String(product.id) },
  }));

  return { paths, fallback: false };
}
```

One quirk of the `paths` generation is that we're converting `product.id` from a number to a string, since the `[id]` path parameter needs to be a string. Next.js would error with `Error: A required parameter (id) was not provided as a string received number in getStaticPaths for /products/[id]` otherwise.

getStaticProps takes the params object, which is contained in the objects returned from getStaticPaths, and makes further fetch calls to load the product by ID. Finally, it returns product for the Page component to use:

```
export async function getStaticProps({ params }) {
  const product = await fetch(
    `https://fakestoreapi.com/products/${params.id}`
  ).then((res) => res.json());
  return {
    props: {
      product,
    },
  };
}
```

Our ProductPage component can then look like the following, where we use product.title both as the title of the page and as the page's h2 element content. From here, we could display anything contained in the product response, including price and stock information and images:

```
import React from 'react';
import Link from 'next/link';
import Head from 'next/head';

export default function ProductPage({ product }) {
  return (
    <>
      <Head>
        <title>{product.title}</title>
      </Head>
      <div>
        <Link href={'/products'}>Back</Link>
        <h2>{product.title}</h2>
      </div>
    </>
  );
}
```

When we run `next build`, the build will take longer, since each `products/[id]` page needs to make a request to `fakestoreapi.com`. Note that the `products/[id]` pages are marked as SSG. We also see the number of static pages being generated increasing to 24 and a truncated subset of `products/[id]` pages:

```
npx next build
info  - Linting and checking validity of types
info  - Creating an optimized production build
info  - Compiled successfully
info  - Collecting page data
info  - Generating static pages (24/24)
info  - Finalizing page optimization


Route (pages)                            Size       First Load JS
┌ ○ /                                    464 B          75.9 kB
├ ○ /404                                 182 B          73.2 kB
├ λ /cart                                445 B          73.5 kB
├ ● /products                            426 B          75.9 kB
└ ● /products/[id]                       383 B          75.9 kB
    ├ /products/1
    ├ /products/2
    ├ /products/3
    └ [+17 more paths]
+ First Load JS shared by all            73.1 kB
  ├ chunks/framework-fcfa81c6fe8caa42.js 45.2 kB
  ├ chunks/main-7039e34bfb6f1a68.js      26.9 kB
  ├ chunks/pages/_app-c7a111f3ee9d686c.js 195 B
  └ chunks/webpack-8fa1640cc84ba8fe.js   750 B


λ  (Server)   server-side renders at runtime (uses getInitialProps or
getServerSideProps)
○  (Static)   automatically rendered as static HTML (uses no initial
props)
●  (SSG)      automatically generated as static HTML + JSON (uses
getStaticProps)
```

After building and starting the server with `next start`, when we load the `/products/1` path, we see product 1's name.

Back

Fjallraven - Foldsack No. 1 Backpack, Fits 15 Laptops

Figure 5.14: /products/1 content

And when we load the `/products/8` path, we see product 8's name.

Back

Pierced Owl Rose Gold Plated Stainless Steel Double

Figure 5.15: /products/8 content

We've now seen how to leverage Next.js features that automatically statically render pages with no data fetching, `getStaticProps` and `getStaticPaths` to render pages with dynamic content and with dynamic paths at build time, as well as how these approaches contrast with `getServerSideProps`.

Next, we'll deep-dive into how to rehydrate a server-rendered react page on the client.

Page hydration strategies

As we've seen in the first section of the chapter, react provides primitives to render applications on the server and the client. However, we only looked at examples where we did exclusively client or server rendering. One key feature of React frameworks such as Next.js is that they allow you to seamlessly switch between static, client, and server rendering. We'll look at how to achieve this using React primitives.

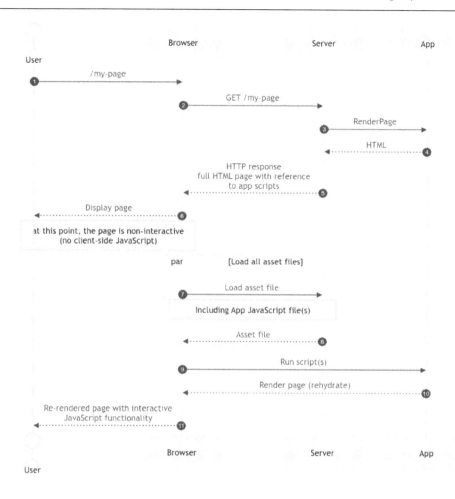

Figure 5.16: A sequence diagram for a server-rendered page that is subsequently rehydrated on the client

We'll start by extending our React client/server rendering app.jsx with a ClientCounter component. Event handlers are one of the simplest ways to observe interactivity primitives. Our ClientCounter component displays a counter that initializes with 0, and on every click of the **Add** button, it increments the count value. We put this component in a src/client-counter.jsx file:

```
import React, { useState } from 'react';

export function ClientCounter() {
  const [count, setCount] = useState(0);
  return (
    <div>
      Dynamic Counter, count: {count}
```

```
        <br />
        <button onClick={() => setCount(count + 1)}>
          Add</button>
      </div>
    );
  }
```

We can render it in our `app.jsx` component, like so:

```
import React from 'react';
import { ClientCounter } from './client-counter';

export function App({ type = '' }) {
  return (
    <>
      <div>
        <p>Hello from the {type + ' '}app</p>
        <ClientCounter />
      </div>
    </>
  );
}
```

If we build the client-side entry point and load it in the browser, it works correctly, incrementing on each **Add** click:

```
npx esbuild client.jsx --bundle --outdir=dist
```

If we open the `index.html` file (which is unchanged), we'll be able to see the counter and increment it, as shown in the following screenshot.

Hello from the "client render" app

Dynamic Counter, count: 7
Add

Figure 5.17: The React client-side rendered counter with an increment of 7 displayed

However, if we build and run our server-side entry point, the component remains at 0:

```
npx esbuild server.js --bundle --platform=node --outdir=dist
```

We can then start the server using the following:

```
node dist/server.js
```

As the following screenshot shows, no matter how many times we click **Add**, the component only ever displays **0**.

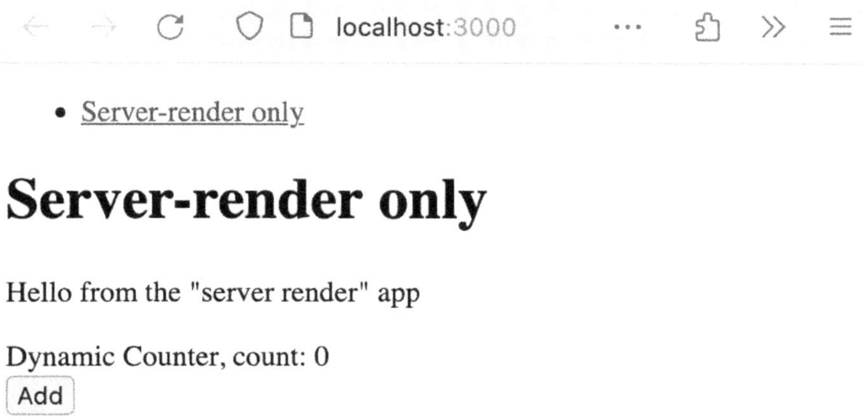

Figure 5.18: React server-rendering does not allow for interactive counter
component, the count shows 0 despite multiple Add button clicks

In order to "hydrate" our server-rendered page, we can create a new entry point, rehydrate.jsx. This uses react-dom/client's hydrateRoot function on the element that contains our app:

```
import React from 'react';
import ReactDOM from 'react-dom/client';
import { App } from './src/app';

ReactDOM.hydrateRoot(
  document.querySelector('#app'),
  <App type={`"server render"`} />
);
```

We'll bundle the rehydration entry point using esbuild, in a similar fashion to previous entry points:

```
npx esbuild rehydrate.jsx --bundle --outdir=dist
```

Once our new `dist/rehydrate.js` file is built, we need to use it in our server-rendered app. We modify `server.js` to statically serve `dist`, which means that `dist/rehydrate.js` is available as `rehydrate.js`. We then create a new GET route, `/rehydrate`. This route returns the navigation elements seen previously, but now the application also has a script that will load `rehydrate.js`:

```
// no changes to other routes
app.use(express.static('./dist'));
app.get('/rehydrate', (_req, res) => {
  res.send(`
    <!DOCTYPE html>
    ${renderNav()}
    <h1>Server-render with client-side rehydration</h1>
    <div id="app">${serverRenderApp()}</div>
    <script src="./rehydrate.js"></script>
  `);
});
// no changes to server startup
```

We also include `/rehydrate` in the `nav.jsx`, which now looks as follows:

```
import React from 'react';
export function Nav() {
  return (
    <ul>
      <li>
        <a href="/">Server-render only</a>
      </li>
      <li>
        <a href="/rehydrate">Server-render with client-side
          rehydration</a>
      </li>
    </ul>
  );
}
```

We can then rebuild our entry points and start the server. When we navigate to `/rehydrate`, the counter is interactive, and we see the navigation and `h1` as rendered server-side.

- Server-render only
- Server-render with client-side rehydration

Server-render with client-side rehydration

Hello from the "server render" app

Rendering: from client

Dynamic Counter, count: 5
[Add]

Figure 5.19: The rehydrated server-rendered application allows the interactive use of a client-side counter, displayed here with a count of 5

We've now seen how to rehydrate a server-rendered React application, next we'll delve into common React rehydration issues.

Common React rehydration issues

Rehydration has some key gotchas.

It's quite common to see the following runtime environment detection code in an application.

```
export const isServer = () => typeof window ===
  'undefined';
```

Let's say we placed isServer in a src/rendering-utils.js file; we can use it as follows to conditionally render content such as 'from client' or 'not from client', or avoid rendering ClientCounter altogether when server-rendering:

```
import React from 'react';
import { ClientCounter } from './client-counter';
import { isServer } from './rendering-utils';

export function App({ type = '' }) {
  return (
    <>
      <div>
```

```
        <p>Hello from the {type + ' '}app</p>
        <p>Rendering: {isServer() ? 'not from client' :
            'from client'}</p>
        {!isServer() && <ClientCounter />}
      </div>
    </>
  );
}
```

This works fine in the purely server-rendered use case, where we display 'not from client' and hide ClientCounter.

Server-render only

Hello from the "server render" app

Rendering: not from client

Figure 5.20: isServer detection working successfully for server-side-only rendering

At first glance, it looks to be working for the server-render followed by client-side rehydration use case. It displays **from client** and shows the client-side counter component.

Server-render with client-side
rehydration

Hello from the "server render" app

Rendering: from client

Dynamic Counter, count: 2
Add

Figure 5.21: isServer detection looking to work for server-side rendering followed rehydration

However, if we look at the console, we can see that we have some errors.

```
⊘ ▷ Warning: Text content did not match. Server: "from client"
    Client: "not from client"
    p
    div
    App@http://localhost:3000/rehydrate.js:35945:15

                                            rehydrate.js:2426:40
⊘ ▷ Warning: An error occurred during hydration. The server HTML
    was replaced with client content in <div>.

                                            rehydrate.js:2426:40
⊘ ▷ Uncaught Error: Text content does not match server-rendered
    HTML.
        checkForUnmatchedText            …st:3000/rehydrate.js:9320
        didNotMatchHydratedTextInstance  …t:3000/rehydrate.js:10593
        prepareToHydrateHostTextInstance …t:3000/rehydrate.js:11465
        completeWork                     …t:3000/rehydrate.js:18213
        completeUnitOfWork               …t:3000/rehydrate.js:21125
```

Figure 5.22: Console errors during rehydration

The issue is client render versus server render mismatches – for example, **Warning: Text content did not match. Server: "from client" Client: "not from client"**. ReactDOM.rehydrateRoot expects the application to render the same way on the server and the client. React, in this situation, falls back to full client-side rendering (**An error occurred during hydration. The server HTML was replaced with client content in <div>.**), meaning the server-rendered HTML is completely thrown away.

To fix this, a better detection of server versus client is required. A simple detection would involve a hook using useEffect. The useClientRenderingOnly function will always be false until the application runs our useEffect, which is only run client-side:

```
export function useClientRenderingOnly() {
  const [hasMounted, setHasMounted] = useState(false);
  useEffect(() => {
    setHasMounted(true);
  });
  return hasMounted;
}
```

It can be used as follows in src/client-counter.jsx instead of isServer in app.jsx:

```
import React, { useState } from 'react';
import { useClientRenderingOnly } from './rendering-utils';


export function ClientCounter() {
```

```
    const isClientRendering = useClientRenderingOnly();
    const [count, setCount] = useState(0);
    if (!isClientRendering) return null;

    // no change to JSX return
}
```

app.jsx can become the following, leveraging isClientRendering to display 'from client' and 'not from client':

```
import React from 'react';
import { ClientCounter } from './client-counter';
import { isClientRendering } from './rendering-utils';

export function App({ type = '' }) {
  return (
    <>
      <div>
        <p>Hello from the {type + ' '}app</p>
        <p>
          Rendering: {isClientRendering ? 'from client' :
            'not from client'}
        </p>
        <ClientCounter />
      </div>
    </>
  );
}
```

In the server-rendering-only case, this works, and in the rehydration case, we now know whether to display something on the server or client without getting rehydration issues.

Other common issues that cause rehydration errors are invalid markup (some HTML tags are not supposed to be inside other HTML tags).

React provides one more rendering approach that allows the server to start returning data to the client earlier via streaming.

React streaming server-side rendering

React streaming server-side rendering leverages streaming so that the server can start return data to the browser earlier (chunks in a stream instead of a one-off response). This also means that the browser can start working on rendering earlier.

There's a major caveat to streaming, which is that one of its key advantages over non-streaming server-rendering is that it has support for the new suspense primitive. This primitive is supported by specific libraries and frameworks and is quite difficult to illustrate using React primitives.

According to the React documentation on suspense usage (`https://react.dev/reference/react/Suspense#usage`):

> *Suspense-enabled data fetching without the use of an opinionated framework is not yet supported. The requirements for implementing a Suspense-enabled data source are unstable and undocumented. An official API for integrating data sources with Suspense will be released in a future version of React.*

When rehydrating a React streaming server rendered page, we need to replace the whole document, so we'll create a new `<Page>` component, which will be a full page. We'll also create a `streaming-rehydrate.jsx` entry point for use client-side.

The following are the contents of a new `src/page.jsx` file. The full page including `html` and `head` are necessary to do streaming server-side rendering:

```
import React from 'react';
import { App } from './app';
import { Nav } from './nav';

export default function Page() {
  return (
    <html>
      <head>
        <title>Streaming</title>
      </head>
      <body>
        <Nav />
        <h1>Server-render with streaming</h1>
        <div id="app">
          <App type={`"streaming server render"`} />
        </div>
      </body>
    </html>
  );
}
```

Our `streaming-rehydrate.jsx` entry point is quite similar to our `rehydrate.jsx` entry point with the exception that it hydrates `document`, instead of an element with the app ID. This is due to the aforementioned limitation of streaming server-side rendering – the whole document has to be controlled by React:

```
import React from 'react';
import ReactDOM from 'react-dom/client';
import Page from './src/page';


ReactDOM.hydrateRoot(document, <Page />);;
```

We'll build the entry point to JavaScript using the following:

npx esbuild streaming-rehydrate.jsx.jsx --bundle --outdir=dist

We can now start working on the server rendering in `src/server-rendering.jsx`. We create a new `serverRenderAppStream` function that takes an Express/Node.js `res` object as a parameter. It calls `ReactDOMServer.renderToPipeableStream` with the `Page` component, and with `bootstrapScripts` set to include our `streaming-rehydrate.js` entry point:

```
import React from 'react';
import ReactDOMServer from 'react-dom/server';
// no changes to other imports
import Page from './page';

export function serverRenderAppStream(res) {
  const { pipe } = ReactDOMServer.renderToPipeableStream
    (<Page />, {
    bootstrapScripts: ['./streaming-rehydrate.js'],
  });
  pipe(res);
}
```

In `server.js`, we can create a new GET route for the `/streaming` path, which simply calls `serverRenderAppStream` with the `res` object per the Express route handler definition:

```
// no change to other imports
import {
  // no change to other imports
  serverRenderAppStream,
} from './src/server-render';

// no change to other routes
app.get('/streaming', (_req, res) => {
  serverRenderAppStream(res);
```

```
});
// no change to startup logic
```

We'll also add the /streaming route to src/nav.jsx:

```
import React from 'react';

export function Nav() {
  return (
    <ul>
      {/* no change to the other li elements */}
      <li>
        <a href="/streaming">Server-render with streaming
          </a>
      </li>
    </ul>
  );
}
```

We can now load the /streaming page and see it in action.

- Server-render only
- Server-render with client-side rehydration
- Server-render with streaming

Server-render with streaming

Hello from the "streaming server render" app

Rendering: from client

Dynamic Counter, count: 3
Add

Figure 5.23: React streaming server rendering with rehydration

We've now seen how to implement React streaming server rendering with rehydration.

Summary

In this chapter, we covered how a deeper understanding of rendering and page hydration strategies can help us deliver optimal and scalable web user interfaces with React.

Client and server rendering have benefits and drawbacks that are complimentary to each other. Client rendering takes longer to start up but provides more interactivity and doesn't require as much server-side computer power; server rendering can return content faster but requires infrastructure and doesn't provide the same level of interactivity.

The static site generation functionality of Next.js can be leveraged alongside classic server rendering to judiciously decide on a rendering strategy for a given set of pages, based on the access pattern and how often the content changes.

Finally, page hydration and rehydration alongside streaming server-side rendering bridges the gap between server and client rendering, allowing the benefits of both to be included in one page.

Now that we're familiar with rendering and page hydration strategies, we can look at implementing micro-frontends using both the "zones" and "islands" architectures in the next chapter.

Micro Frontends, Zones, and Islands Architectures

The micro frontend architecture, and specifically the "zones" and "islands" patterns, mirror the microservices architecture for backend systems. Given the right tooling, they allow multiple teams to maintain high-velocity development on a single product. The techniques covered in this chapter look at system-level interaction and integration patterns. Each system can leverage creational, structural, behavioral, and reactive view library patterns, as covered in *Chapters 1, 2, 3*, and *4* respectively. Micro frontend architectures help link systems together as opposed to structuring the code within each of them better.

We'll cover the following topics in this chapter:

- The problem space that micro frontends address, including some common approaches and their drawbacks
- Leveraging Next.js features to build a "zones" micro frontend setup
- Using the `is-land` package to deliver an "islands" micro frontend setup with islands in Preact and Vue.js

By the end of this chapter, you'll be able to discuss the trade-offs and deliver modern micro frontend approaches in JavaScript.

Technical requirements

You can find the code files for this chapter on GitHub at `https://github.com/PacktPublishing/Javascript-Design-Patterns`

An overview of micro frontends

A micro frontend setup is one where multiple frontend applications or components are composed. This is akin to microservices, where a micro frontend would encapsulate a subset of functionality, or "bounded context."

For example, in an e-commerce setting, we might have a "search" micro frontend and a "cart" or "checkout" micro frontend.

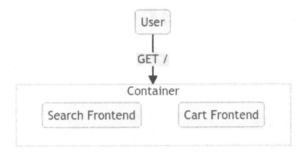

Figure 6.1: A simplified micro frontends diagram

We've now introduced the micro frontends architecture; next, we'll see at the key benefits micro frontends deliver.

Key benefits

The benefits of the micro frontend patterns are similar to microservice benefits. They usually come in the sociotechnical aspect of development.

Each micro frontend can use a different set of technologies, which means the right tool for the job can be selected. A very page-load performance-sensitive page might use a different stack than an admin interface or a high-volume SVG visualization page.

Incremental upgrades are available, and changes can be tested in one component before being rolled out to all components.

The releases of different micro frontends are not locked together. This can help when scaling, where each team might work on one or more of the micro frontends. They can be released independently of other teams, meaning the cadence can increase; this is related to the last benefit we'll discuss.

Each micro frontend can have its own code base, and "bounded contexts" can be strictly enforced.

"Classic" micro frontend patterns

We'll cover five different "classic" approaches to creating a micro frontend setup.

The first is the "container application" using server-side includes. This leverages a server that will fetch from the different micro frontends and stitch them together. This is illustrated in the following diagram, where the container application loads a "cart" HTML section and a "search" HTML section and injects them into its own template, before returning to the client.

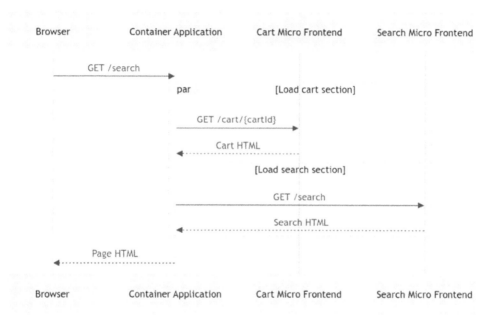

Figure 6.2: The "container application" sequence

The benefits of server-side include, or the "container application," is that deployments of each micro frontend are decoupled (e.g., we can deploy changes to the cart without deploying changes to the search section or the container); in addition, it's completely technology-agnostic, as micro frontends don't even need to use JavaScript.

The next "classic" approach we'll see is different, using "build-time composition," where each micro frontend is a package, usually a npm package (part of the Node.js/JavaScript toolchain). Each package is then imported where necessary and composed at "build-time" (when each application is packaged for deployment).

The key drawback of "build-time" composition is that releases now require deployment cascades. For all the applications to receive updates to the "cart", we need to update the version and release all the applications.

The three final "classic" approaches are similar conceptually although use different technologies. They're all "runtime integrations"; the technologies are iframes, JavaScript, and web components. Runtime integrations mean that the micro frontend requests the micro frontend resource from the browser.

In the case of iframes, this involves using the iframe `src` attribute. The main downside of this is that each micro frontend needs to be secured against all public network exploits. What's more, allowing the iframing of an application's content can lead to click-jacking vulnerabilities if not done carefully, so there are security implications.

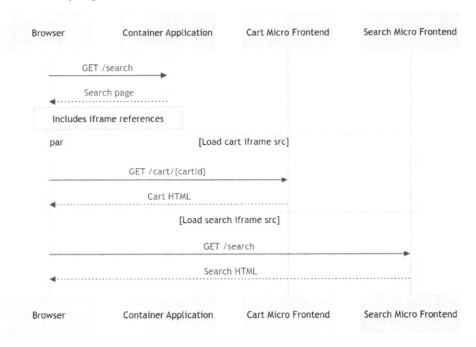

Figure 6.3: Runtime integration with iframes

In the case of JavaScript or Web Component "runtime" integration, the composition is managed by loading JavaScript files. This is more ideal than using iframes, since serving JavaScript to the browser has fewer security implications than allowing the framing of your content. In the Web Components case, you would have both the web component referenced in the body of the HTML and a reference to the scripts required to run the Web Component.

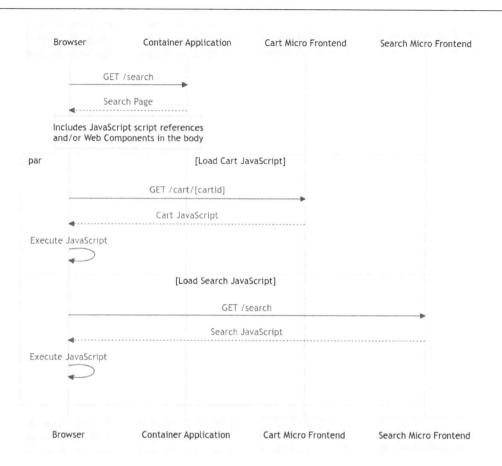

Figure 6.4: Runtime integration with JavaScript or web components

Runtime integrations have a performance impact on the user experience, as shown by the difference in diagrams between our "server-side includes" and "runtime integration" diagrams. In the "server-side includes" case, the server composes a full application before returning it to the customer. In the "runtime integration", the server returns what are essentially resource references to the browser, which then has to load these resources.

As we explore the "zones" and "islands" modern implementations of micro frontends, we'll encounter several instances of these techniques used together.

Other concerns in a micro frontend world

Similar to a microservice setup, micro frontends that allow different teams to build in their own way can be a benefit and a drawback.

At the end of the day, most frontend systems will need to communicate with a backend service. How to do this remains something to be decided – should each team deploy its own **backend for frontend** (**BFF**), should a single gateway be deployed that exposes relevant service endpoints, or should it be a gateway that wraps services in a different query system such as GraphQL?

Micro frontends also cause challenges for testing. How do we reliably test at the "user journey" level, which might go through multiple micro frontends, when each of the micro frontends also has its own test suite?

Similar to the questions about which backend integrations to use, there's a challenge related to shared styles and potentially component libraries. Teams doing micro frontends might standardize on a set technology (React, Vue, etc) in order to gain the benefits of a component library. Component libraries are more difficult to maintain in multiple technologies, but companies sometimes opt for this to support their engineers in picking the right tool for the job.

One big challenge of micro frontends is how to keep them performant. Even in a case where all teams use the same technologies, it's likely that the same dependency is duplicated across micro frontends, which has a performance impact. When technologies and build and deploy processes diverge (which is possible with micro frontends), this problem is exacerbated.

The other performance issue that occurs with, for example, "server-side includes" is that the page will only load as fast as the slowest component on the page. This is less of an issue with runtime integrations, but the idea that each micro frontend might affect a whole page's performance is a relevant one with regard to the challenges of building a system using micro frontends.

Finally, as we've alluded to with regard to testing micro frontends, it causes operational and governance complexities. For example, environment mismatch issues are harder to detect. Running or deploying a full environment for development or testing with multiple micro frontends is more complex than with a monolithic application.

Now that we've defined and contrasted the benefits and drawbacks of micro frontends in general and specific micro frontend approaches, we can look at modern implementations of micro frontends. In the following section, we'll look at leveraging Next.js and "zones" in order to build flexible micro frontends.

Composing applications with Next.js "zones"

Next.js "zones" are a URL "base path"-driven approach to composing Next.js applications. This allows us to build a micro frontend setup with Next.js.

What this means, as shown in the figure that follows, is that an e-commerce use case, where the user might request four sets of URLs (GET /, GET /careers, GET /search, and GET /cart/{id}), "{id}"), denotes that the cart has a dynamic segment, which is the cart ID that is requested. For GET / and GET /careers, the request first goes to the root frontend, which handles rendering directly. For GET /search, the request goes to the root frontend, which forwards the request to the search frontend. Similarly, for GET /cart/{id} requests, the request initially is sent to the root frontend, which proxies the request to the checkout frontend.

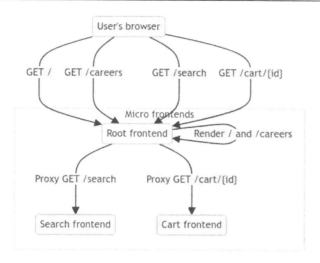

Figure 6.5: An overview flowchart of a three-app Next.js zone setup

We've now introduced Next.js "zones" and an overview of our implementation, next we'll implement the "root app".

Root app

The root app contains two pages, / (pages/index.js) and /careers (pages/careers. js). Both of these pages are statically rendered, index.js via automatic static generation (since it doesn't have getServerSideProps or getInitialProps) and careers.js via static site generation (since it has getStaticProps).

index.js contains a heading as well as Head content.

```
import React from 'react';
import Head from 'next/head';

export default function Home() {
  return (
    <>
      <Head>
        <title>Homepage (Root zone)</title>
      </Head>
      <main>
        <h1>Root</h1>
      </main>
    </>
  );
}
```

When we load the GET / path, our root app renders the h1 element we placed within it, with 'Root' as its content.

Root

Figure 6.6: The Root page rendered

The /careers page loads roles from an API using getStaticProps and displays them in a list.

We can start with a getStaticProps function in pages/careers.js. This function loads from a "fake jobs" API and returns a roles prop, which includes the returned data from the API:

```
export async function getStaticProps() {
  const jobs = await fetch(
    'https://apis.camillerakoto.fr/fakejobs/
      jobs?fulltime=true').then((res) => res.json());
  return {
    props: { roles: jobs },
  };
}
```

Next, we'll add a CareersPages component. It includes the page scaffolding with title and h1. It also loops through the roles prop to render it as a list, using ul and li:

```
import React from 'react';
import Head from 'next/head';

export default function CareersPage({ roles }) {
  return (
    <>
      <Head>
        <title>Careers (Root zone)</title>
      </Head>
      <main>
        <h1>Careers</h1>
        <ul>
          {roles.map((role) => {
            return (
              <li key={role.id}>
                {role.title} ({role.country})
              </li>
```

```
            );
        })}
        </ul>
      </main>
    </>
  );
}
```

It displays as follows.

Careers

Professional Bus Driver (China)

Video game creator (Turkey)

Software Engineer (United States of America)

Web Developer (FullStack) (Japan)

Professional seller (China)

Product engineer and designer (France)

Figure 6.7: The Careers page in the root zone

The next build output shows that index.js is indeed statically rendered, while /careers uses static site generation:

```
Route (pages)                               Size      First Load JS
┌ ○ /                                        430 B         77.7 kB
├ ○ /404                                      182 B         77.5 kB
├ λ /api/health                               0 B           77.3 kB
└ ● /careers                                  498 B         77.8 kB
+ First Load JS shared by all               77.3 kB
  ├ chunks/framework-4725d5bb117f1d8e.js     45.2 kB
  ├ chunks/main-7a398668474d4dd1.js          31.1 kB
  ├ chunks/pages/_app-ecd5712b2c05cb6a.js    195 B
  └ chunks/webpack-8fa1640cc84ba8fe.js       750 B

λ  (Server)  server-side renders at runtime (uses getInitialProps or
getServerSideProps)
○  (Static)  automatically rendered as static HTML (uses no initial
props)
●  (SSG)     automatically generated as static HTML + JSON (uses
getStaticProps)
```

We've now started implementing the root app, we'll move on to our second zone, the "search" zone.

Adding a /search app

Next, we'll build and mount a /search page.

search/pages/index.js displays an input and makes a call to the /search/api/search route on change:

```
import React, { useState } from 'react';
import Head from 'next/head';

export default function Home() {
  const [searchResult, setSearchResult] = useState({
    count: 0,
    matches: [],
  });
  return (
    <>
      <Head>
        <title>Search Page (Search zone)</title>
      </Head>
      <main>
        <h1>Search</h1>
        <input
          type="search"
          onChange={async (event) => {
            const data = await fetch(
              `/search/api/search?q=${event.target.value}`
            ).then((res) => res.json());
            setSearchResult(data);
          }}
        />

        <div>
          <h2>Results ({searchResult.count})</h2>
          {searchResult.matches.map((product) => {
            return <div key={product.id}>
              {product.title}</div>;
          })}
        </div>
      </main>
```

```
      </>
   );
}
```

To implement the `search/pages/api/search` API route, we create `pages/api/search`, which loads products from `fakestoreapi` and finds a match between the title, description and category:

```
export default async function handler(req, res) {
   const allProducts = await fetch
      ('https://fakestoreapi.com/products').then(
         (res) => res.json()
   );
   const { q } = req.query;
   const searchQuery = Array.isArray(q) ? q[0] : q;
   const matches = allProducts.filter(
      (product) =>
         product.title.includes(searchQuery) ||
         product.description.includes(searchQuery) ||
         product.category.includes(searchQuery)
   );
   return res.status(200).json({ matches,
      count: matches.length });
}
```

In order for `search-app/` to be mounted under `search-app/search`, we'll use `basePath` in `next.config.js` in the search app:

```
module.exports = {
   basePath: '/search',
};
```

We'll expose `search` via the root app's `next.config.js`:

```
module.exports = {
   async rewrites() {
      return [
         {
            source: '/search/:path*',
            destination: 'http://localhost:3001/search/:path*',
         },
      ];
   },
};
```

We can then load the **Search** page, which displays as follows:

Figure 6.8: The Search page on load

The search works – for example, with the jacket search term.

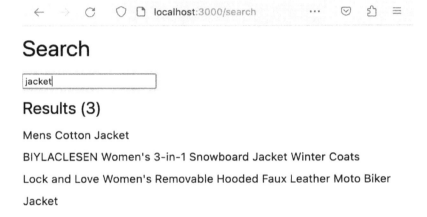

Figure 6.9: The Search page with the jacket search term

/search is statically rendered via automatic static rendering:

```
Route (pages)                                Size        First Load JS
┌ ○ /                                        607 B          73.7 kB
├   └ css/776983a5dfcef528.css               271 B
├ ○ /404                                     182 B          73.2 kB
├ λ /api/health                              0 B            73.1 kB
└ λ /api/search                              0 B            73.1 kB
+ First Load JS shared by all                73.1 kB
  ├ chunks/framework-4725d5bb117f1d8e.js     45.2 kB
  ├ chunks/main-ee0b7fc0f7162449.js          26.9 kB
  ├ chunks/pages/_app-ecd5712b2c05cb6a.js    195 B
  └ chunks/webpack-ab5c478f511867a3.js       756 B
```

```
λ  (Server)   server-side renders at runtime (uses getInitialProps or
getServerSideProps)
○  (Static)   automatically rendered as static HTML (uses no initial
props)
```

We've now implemented the search "zone", next we'll implement the checkout "zone".

Adding /checkout app

We'll add a "view cart" page in a new checkout Next.js app under `pages/cart/[id].js`.

The cart page loads a cart and its contained products from `fakestoreapi`, displaying them via a `CartContents` component.

First, we'll define a `CartContents` component that takes `cart` and `productsById` props. It then maps through `cart.products`, extracting the product's title and the quantity requested in the cart, before computing and formatting the price in euros using `.toLocaleString`.

The reason we need `cart` and `productsById` is that the cart comes back in a normalized data format, meaning it contains only cart-specific information and none of the related product's information, except the product ID. Therefore, we need to do a lookup based on the product ID.

Our rendering logic uses an unordered list container (the `ul` HTML element) and list item elements (`li` HTML elements). We render the title in an `h3` heading and the rest of the information using `span` elements:

```
import React from 'react';

function CartContents(props) {
  const { cart, productsById } = props;
  return (
    <ul>
      {cart.products.map((product) => {
        const fullProductInformation = productsById
          [product.productId];
        return (
          <li key={product.productId} className=
            "cart-item-product">
            <h3 className="cart-item-product-name">
              {fullProductInformation?.title}
            </h3>
            <span className="cart-item-product-quantity">
              {' '}
              x {product.quantity}
            </span>
            <span className="cart-item-product-price">
```

```
                    Price:
                    {(
                       product.quantity *
                         fullProductInformation?.price)
                         .toLocaleString('en', {
                       style: 'currency',
                       currency: 'EUR',
                    })}
                  </span>
                </li>
             );
          })}
        </ul>
     );
  }
```

Now that we're rendering the contents of the cart, the additional functionality we'll add to CartContents is a display of the cart's total price.

This is done by adding another li, which displays "Total:" and computes the total price, using reduce over cart.products. Remember from the previous code block that cart.products is normalized, meaning that it doesn't contain any information about the product (e.g., its price), apart from the product's ID. This means that our reduce handler does a lookup on productsById[product.productId] in order to access the product's price.

Once we have the quantity of a given product in the cart and the price of the product, we simply multiply them together and sum the quantity times the price result to the accumulator, which we initialized as 0.

Similar to the cart items, we use toLocaleString to format the total price in euros as an en localized string:

```
// no change to imports

function CartContents(props) {
  // no change to the function body
  return (
    <ul>
      {/* no change to `cart.products` mapping */}
      <li className="cart-item-product">
        <strong className="cart-item-product-price">
          Total:
          {cart.products
            .reduce((acc, curr) => {
              const fullProductInformation = productsById
```

```
                    [curr.productId];
                  return acc + curr.quantity *
                    fullProductInformation.price;
                }, 0)
                .toLocaleString('en', {
                  style: 'currency',
                  currency: 'EUR',
                })}
          </strong>
        </li>
      </ul>
    );
  }
```

We'll leverage `getServerSideProps` to load the cart, and then the relevant products from `fakestoreapi`. As mentioned in the previous code blocks, `fakestoreapi`'s cart response is normalized and, therefore, doesn't include all the product data we need, which is why we load the products by ID.

Once we have a cart response and all relevant product responses, we process the products to allow them to be looked up by ID. Finally, `getServerSideProps` returns `id` (the cart ID from the Next.js context), `productsById`, and `cart` in a `props` property of an object so that Next.js can pass them to our page component:

```
// no changes to imports
// no changes to CartContents definition
export async function getServerSideProps(ctx) {
  const { params } = ctx;
  const cartId = params.id;
  const cart = await fetch(`https://fakestoreapi.com/carts
    /${cartId}`).then(
    (res) => res.json()
  );
  if (!cart?.products) {
    return {
      props: {
        id: cartId,
      },
    };
  }
  const productsById = (
    await Promise.all(
      cart.products.map(async (product) => {
        return await fetch(
```

```
            `https://fakestoreapi.com/products/$
                {product.productId}`
          ).then((res) => res.json());
        })
      )
    ).reduce((acc, curr) => {
      acc[curr.id] = curr;
      return acc;
    }, {});
    return {
      props: {
        id: cartId,
        cart,
        productsById,
      },
    };
}
```

Finally, we'll add our `GetCartPage` component, which will take props as passed by Next.js (based on the output of `getServerSideProps`), and we'll use them to render `CartContents`, as well as a heading and title:

```
import Head from 'next/head';
import React from 'react';

// no changes to CartContents definition

export default function GetCartPage({ id, cart, productsById }) {
  return (
    <>
      <Head>
        <title>GetCartPage (Checkout zone)</title>
      </Head>
      <main>
        <h1>GetCartPage (Checkout zone)</h1>
        <CartContents cart={cart} productsById=
          {productsById} />
      </main>
    </>
  );
}

// no changes to getServerSideProps definition
```

For the checkout app to mount under the right path, we set `basePath` in its `next.config.js`:

```
module.exports = {
  basePath: '/checkout',
};
```

We again need to modify the root app's `next.config.js` so that relevant requests are proxied to the checkout app:

```
module.exports = {
  async rewrites() {
    return [
      // no change to other entries
      {
        source: '/checkout/:path*',
        destination:'http://localhost:3002/checkout/:path*',
      },
    ];
  },
};
```

We can load `/checkout/cart/2`, and the following will display:

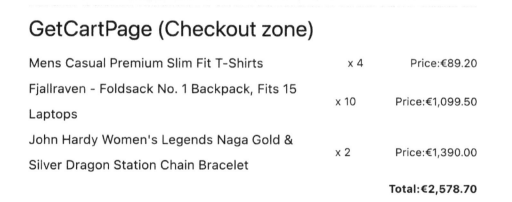

Figure 6.10: The cart/[id] page in the checkout zone with cart 2 loaded

In the build output, we can see that `/cart/[id]` is server-side rendered, since it uses `getServerSideProps`:

```
Route (pages)                                Size      First Load JS
┌ ○ /                                        445 B            73.5 kB
├ ○ /404                                      182 B            73.3 kB
└ λ /cart/[id]                               3.95 kB            77 kB

λ  (Server)  server-side renders at runtime (uses getInitialProps or
getServerSideProps)
○  (Static)  automatically rendered as static HTML (uses no initial
props)
```

We've now seen how to add a checkout "zone" to our micro frontend setup, next we'll cover the benefits of a micro frontend "zones" architecture specifically with regards to working in growing teams.

The benefits/supporting team scaling

Using "zones" with `basePath` means that Next.js features work out of the box. For example, client-side transitions and `getServerSideProps` re-fetches work (where Next.js loads `{basePath}/_next/...`) as well as the API routes that we used in the search example.

Adding new pages also "just works"; a new page at `/cart/[id]/checkout` wouldn't require any changes to the root app to be available to users.

The only time we would change the root application config is to add a whole new app (top-level path) – for example, if we wanted an admin app, we would need to create that and configure the root `next.config.js`.

In the case where there's a lot of traffic to applications and we want to be more efficient with our resource, we don't need to use the `root` app to forward all requests to the other micro frontends; we could leverage any reverse proxy (web servers such as NGINX and Caddy) or even the infrastructure provider's CDN (e.g., Fastly, Akamai, Cloudflare and AWS) can be configured to forward all requests from `domain.tld/{path}/*` (all requests to `domain.tld` starting with `{path}`) to a specific origin.

By having a suite of applications that all use Next.js, pages can be built in the root app experimentally and then spun off to a full Next.js application.

For debugging and communication purposes, having the app name in the URL can help when discussing apps and pages with technical and non-technical team members. For example, even non-technical team members will understand that "this first section of the URL is the application name."

One other benefit of using zones is that that request is not rewritten during a reverse-proxy pass. For example, in some setups, the reverse proxy would receive `/search` but load `/` on the search app. This means that there's a subtle mismatch when running the search app locally versus when proxied.

Next.js was used for all systems here but is not required; most tools can be configured to serve out of a "sub-path," or `basePath` as Next.js calls it.

The drawbacks of Next.js zones

In the setup we've demonstrated, the key drawback is that the "framework" bundle is not shared across apps. That means that as a user goes from one zone to another, they load a different version of Next.js, React, and React DOM. This is suboptimal but probably acceptable for a lot of use cases. When it's not acceptable, a technique such as module federation or its predecessor, vendor bundles, can be deployed.

Another drawback is that when developing locally, using `next dev`, and using the root app to proxy requests, we lose features such as fast refresh/live reload. This can be worked around by going directly to the micro frontend during local development.

Now that we understand how we can use Next.js path-based routing, proxying, and base URL functionality to deliver a "zones" implementation, where the micro frontends each serve different subsets of the URLs, we'll now look at how to deliver a micro frontend application where all the micro frontends are visible on one page, using the "islands" architecture with the `is-land` package. The micro frontends will be built using Preact and Vue.js.

Scaling performance-sensitive pages with the "islands" architecture

According to the is-land library documentation (`https://github.com/11ty/is-land`), is-land is *"A new performance-focused way to add interactive client-side components to your web site. Or, more technically: a framework independent partial hydration islands architecture implementation."*

Let's start by looking at what a "islands architecture" is. The islands architecture is a paradigm where a page is mainly server-rendered, and interactivity is added specifically where necessary. This reduces the page load time, as well as the amount of JavaScript being delivered (JavaScript is only delivered for specific client-side interactions). This is in contrast to situations where a JavaScript application "takes over" the full page – for example, in a Next.js app, where the client-side JavaScript will remount what's been server-rendered, meaning the minimum amount of JavaScript running client-side by default is Next.js client code + React + React DOM.

The following diagram shows how the islands architecture can be leveraged to deliver a micro frontend experience.

Each island is responsible for its own data fetching from the server.

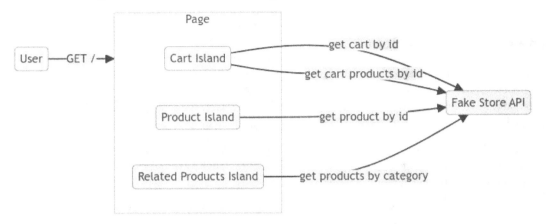

Figure 6.11: An app page composed of islands

One additional element in the islands architecture is loading JavaScript on user interaction – for example, on a click, hover, or scroll into view of an element. The is-land package provides primitives to create islands with these types of hydration strategies.

Islands setup with is-land

We'll look at how to implement the three-islands page with a product island that is immediately initialized, a cart island that is initialized on interaction, and a related products island that is initialized when scrolled into view.

Our example will demonstrate the usage of all the tools without a bundler in the first instance. We'll use Preact with *htm* (since we don't have a JSX compilation pipeline) and Vue with DOM templates.

In order to enable simple imports in our scripts, we'll leverage an import map, loading from the unpkg.com CDN:

```
<script type="importmap">
  {
    "imports": {"@11ty/is-land/is-land.js":
      "https://unpkg.com/@11ty/is-land@4.0.0/is-land.js",
      "htm/preact": "https://unpkg.com/htm@3.1.1/
        preact/index.module.js",
      "htm": "https://unpkg.com/htm@3.1.1/dist/htm.mjs",
      "preact": "https://unpkg.com/preact@10.15.1/
        dist/preact.mjs",
      "vue": "https://unpkg.com/vue@3.2.36/dist
        /vue.esm-browser.prod.js"
```

```
    }
  }
</script>
```

In order to initialize the islands, we'll include the is-land package at the end of the page:

```
<script type="module">
  import '@11ty/is-land/is-land.js';
</script>
```

We've now introduced the page we'll be building and configured is-land to initialize on page load, next we'll implement the product island.

Product island

We'll use Vue to build our product island.

The first step is to create an <is-land> element and script.

We set on:visible so that the contents of the island are initialized by is-land when the element is within the viewport; since our HTML contains only the product island, this will occur on page load.

We'll create a Vue app that, on mount, makes an API call to fakestoreapi.com to fetch a product based on the query parameters. Around the API call using fetch, we'll set this.loading = true (before the API call starts) and this.loading = false (when the API call completes).

The data method of the Vue app will read productId from the URL query string, set loading to true, and set product to an empty object literal ({ }):

```
<is-land on:visible>
  <div id="vue-product-island"></div>

  <template data-island>
    <script type="module">
      import { createApp } from 'vue';

      createApp({
        async mounted() {
          this.loading = true;
          const product = await fetch(
            `https://fakestoreapi.com/
              products/${this.productId}`
          ).then((res) => res.json());
          this.product = product;
          this.loading = false;
        },
```

```
        data: () => ({
          productId:
            new URLSearchParams(window.location.search).
              get('productId') || '1',
          loading: true,
          product: {},
        }),
      }).mount('#vue-product-island');
    </script>
  </template>
</is-land>
```

Now that the data is loaded, we can focus on the template; we'll render the title, description, and other product information:

```
<is-land on:visible>
  <div id="vue-product-island" class="product-container">
    <h2 v-text="product.title"></h2>
    <p v-text="product.description"></p>

    <p v-cloak>
      <span
        v-text="product.price?.toLocaleString('en', {
          style: 'currency', currency: 'EUR'})"
      ></span>
      <br /><span v-text="product?.rating?.rate">
        </span>/5.0 (<span
        v-text="product?.rating?.count"
      ></span>
      >)
    </p>
    <img v-bind:src="product.image" width="320px"
      class="product-image" />
  </div>

  <style>
    .product-container {
      min-height: 100vh;
      border-bottom: solid 1px black;
    }
    [v-cloak] {
      display: none;
    }
  </style>
```

```
  <template data-island>
    <style>
      .product-image {
        min-width: 320px;
        display: block;
        margin: auto;
      }
    </style>
    <!-- no change to the script -->
  </template>
</is-land>
```

When we load this page with `productId=1` or no `productId` (since it's defaulted), we see the following output:

Figure 6.12: The product with ID 1 displaying in the product island

We've now seen how to implement a product island using `is-land` and Vue. Next, we'll build the cart island.

Cart island

Again, we'll start with an `is-land` element, this time with `on:interaction`, which means the island will initialize only when the user clicks on it (we'll show a button for them to do so):

```
<is-land on:interaction>
  <div id="preact-cart-island">
    <button>My Cart</button>
  </div>
</is-land>
```

Next, we'll build a `CartContainer` component that will be mounted using Preact.

`CartContainer` loads cart and product information from `fakestoreapi.com` and stores it in state for a `CartContents` component to render:

```
<is-land on:interaction>
  <div id="preact-cart-island">
    <button>My Cart</button>
  </div>
  <template data-island>
    <script type="module">
      import { html, render } from 'htm/preact';
      import { useState, useEffect } from 'preact/hooks';

      function CartContents() {
        // empty for now
        return null;
      }

      function CartContainer(props) {
        const cartId = props.id ?? 1;
        const [open, setOpen] = useState(true);
        const [isLoading, setIsLoading] = useState(false);
        const [cartContents, setCartContents] = useState({
          cart: null,
          productsById: null,
        });
        useEffect(async () => {
          setIsLoading(true);
          const cart = await fetch(
            `https://fakestoreapi.com/carts/${cartId}`
          ).then((res) => res.json());
          if (!cart?.products) {
            return {
```

```
      props: {
        id: cartId,
      },
    };
  }
  const productsById = (
    await Promise.all(
      cart.products.map(async (product) => {
        return await fetch(
          `https://fakestoreapi.com/
            products/${product.productId}`
        ).then((res) => res.json());
      })
    )
  ).reduce((acc, curr) => {
    acc[curr.id] = curr;
    return acc;
  }, {});

  setCartContents({
    cart,
    productsById,
  });
  setIsLoading(false);
}, [cartId]);

const cartItemCount = cartContents?.
  cart?.products?.length;

return html`<div>
  <button onClick=${() => setOpen(!open)}>
    My Cart ${cartItemCount !== undefined ? `
      (${cartItemCount})` : ''}
  </button>
  ${open && isLoading && html`<div>
    Loading...</div>`} ${open &&
  !isLoading &&
  cartContents.cart &&
  cartContents.productsById &&
  html`<${CartContents}
    cart=${cartContents.cart}
    productsById=${cartContents.productsById}
  />`}
```

```
      </div>`;
    }

    const appContainer = document.querySelector
      ('#preact-cart-island');
    render(
      html`<${CartContainer}
        id=${new URLSearchParams(window.location.search)
          .get('cartId')}
      />`,
      appContainer,
      appContainer
    );
  </script>
  </template>
</is-land>
```

Finally, we'll implement `CartContents`, in which we loop through the cart and render pricing information:

```
<template data-island>
  <script type="module">
    import { html, render } from 'htm/preact';
    // no changes to imports

    function CartContents(props) {
      const { cart, productsById } = props;
      return html`<ul>
        ${cart.products.map((product) => {
          const lineItemQueryParams = new URLSearchParams([
            ['productId', product.productId],
            ['cartId', cart.id],
          ]);
          const fullProductInformation = productsById
            [product.productId];
          return html`<li class="cart-item-product"
            key=${product.productId}>
            ${html`<a href=${'?' +
              lineItemQueryParams.toString()}
              >${fullProductInformation?.title}</a
            >`}
            <span class="cart-item-product-quantity"
              >x ${product.quantity}</span
            >
```

```
              <span class="cart-item-product-price">
                Price:${' '}${(
                  product.quantity * fullProductInformation
                  ?.price).toLocaleString(navigator.language, {
                    style: 'currency',
                    currency: 'EUR',
                  })}
              </span>
            </li>`;
          })}
          <li class="cart-item-product">
            <strong class="cart-item-product-price">
              Total:${' '} ${cart.products
                .reduce((acc, curr) => {
                  const fullProductInformation =
                      productsById[curr.productId];
                  return acc + curr.quantity *
                    fullProductInformation.price;
                }, 0)
                .toLocaleString(navigator.language, {
                  style: 'currency',
                  currency: 'EUR',
                })}
            </strong>
          </li>
        </ul> `;
      }
      // no changes to CartContainer component
    </script>
  </template>
```

When we load our page with cartId 1 and productId 1 and open the cart contents, we can see that it renders the cart with ID 1, including the three line items, their amount, the subtotal per item, and the cart total.

Figure 6.13: Cart 1 rendering in the cart island

We've now implemented the cart island with Preact, next we'll implement a related products island which initializes only when it's visible.

A related products island

Finally, we'll build our related products island. The island itself is quite straightforward, but communicating which product is being displayed and its category is trickier.

We'll build an island that waits to be visible to initialize itself, again using on:visible but also on:idle. This means the island will load either when it's visible or when other processing has completed.

The island will mount if it receives a product-category-load custom event.

We'll start by building the RelatedProducts component, which will receive three props – selectedProductId, category, and from. The from value will be displayed in an h3 element we render to illustrate how the island received its data:

```
<is-land on:visible on:idle id="related-products-island-wrapper">
  <template data-island="">
    <script type="module">
      import { html } from 'htm/preact';

      function RelatedProducts({ selectedProductId,
        category, from }) {
        return html`<div>
          <h3>Related Products (from ${from})</h3>
        </div>`;
      }
```

```
    </script>
  </template>
</is-land>
```

Next, based on the category, we want to load all possible products from `fakestoreapi.com`. We'll store the value using the `useState()` hook, and loading the related products will be done on component mount, using the `useEffect()` hook.

The data fetching logic is as follows. We'll make an API call to `fakestoreapi.com` using the provided category. To fulfill the "related" requirement of the "related products," we'll exclude the product that's currently being displayed – that is, remove the product with an ID equal to `selectedProductId` from the products list. Finally, we sort the related products by rating and persist the first three items to state, using `setRelatedProducts`:

```
<is-land on:visible on:idle id="related-products-island-wrapper">
  <template data-island="">
    <script type="module">
      // no changes to other imports
      import { useState, useEffect } from 'preact/hooks';

      function RelatedProducts({ selectedProductId,
        category, from }) {
        const [relatedProducts, setRelatedProducts] =
          useState([]);
        useEffect(async () => {
          const productsInCategory = await fetch(
            `https://fakestoreapi.com/products/category/$
              {encodeURIComponent(
              category
            )}`
          ).then((res) => res.json());

          const topRelatedProductsByRating =
            productsInCategory
            .filter((el) => {
              return el.id !== parseInt(selectedProductId,
                10);
            })
            .sort((a, b) => b.rating.rate - a.rating.rate);

          setRelatedProducts
            (topRelatedProductsByRating.slice(0, 3));
        }, [selectedProductId, category]);
```

```
                // no change to returned template
            }
        </script>
    </template>
</is-land>
```

With the data persisted to `relatedProducts`, we can now render them using the `.map` function which returns a list. For each product, we want to show a title that's also a link to view the product, its price, an image, and the rating information:

```
<is-land on:visible on:idle id="related-products-island-wrapper">
    <template data-island="">
        <script type="module">
            // no changes to imports

            function RelatedProducts({ selectedProductId,
                category, from }) {
                const [relatedProducts, setRelatedProducts] =
                    useState([]);
                // no change to useEffect

                return html`<div>
                    <h3>Related Products (from ${from})</h3>
                    <ul class="related-product-card-row">
                        ${relatedProducts.map((product) => {
                            const productSearchParams = new
                                URLSearchParams([
                                ['productId', product.id],
                            ]);
                            const currentCartId = new URLSearchParams(
                                window.location.search
                            ).get('cartId');
                            if (currentCartId) {
                                productSearchParams.set('cartId',
                                    currentCartId);
                            }
                            return html`<li class="related-product-card">
                                <a href=${'?' + productSearchParams
                                .toString()}>
                                <h4>${product.title}</h4>
                                <p>
                                    ${product.price.toLocaleString
                                        (navigator.language, {
                                        style: 'currency',
```

```
                        currency: 'EUR',
                      })}
                  </p>
                  <img height="100px" src=${product.image} />
                  <p>${product.rating.rate}/5.0
                      (${product.rating.count})</p>
                </a>
              </li>`;
            })}
          </ul>
        </div>`;
      }
    </script>
  </template>
</is-land>
```

Finally, we'll add logic to mount `RelatedProducts`, based on an event listener for the `product-category-load` custom event:

```
<is-land on:visible on:idle id="related-products-island-wrapper">
  <div id="preact-related-products-island">
    <h3>Related Products</h3>
    <div class="related-product-card-row">Loading...</div>
  </div>
  <template data-island="">
    <script type="module">
      import { html, render } from 'htm/preact';
      // no change to preact/hooks import or
          RelatedProducts

      const relatedProductsIslandContainer =
        document.querySelector(
        '#preact-related-products-island'
      );
      function mountRelatedProductsIsland(
        relatedProductsIslandContainer,
        category,
        selectedProductId,
        from
      ) {
        if (category && selectedProductId) {
          render(
            html`<${RelatedProducts}
              category=${category}
```

```
                    selectedProductId=${selectedProductId}
                    from=${from}
                />`,
                relatedProductsIslandContainer,
                relatedProductsIslandContainer
            );
        }
    }

    document.addEventListener('product-category-load',
        (event) => {
        const category = event.detail.category;
        const selectedProductId = event.detail.
            selectedProductId;

        mountRelatedProductsIsland(
            relatedProductsIslandContainer,
            category,
            selectedProductId,
            'custom-event'
        );
    });
    </script>
  </template>
</is-land>
```

Now, we need to ensure that `product-category-load` is dispatched from the product island. We need to make the following change to the "mounted" life cycle hook of the Vue.js product island script:

```
<script type="module">
  import { createApp } from 'vue';

  createApp({
    async mounted() {
      // no changes
      document.dispatchEvent(
        new CustomEvent('product-category-load', {
          detail: {
            category: this.product.category,
            selectedProductId: this.product.id,
          },
        })
      );
    },
```

```
    // no changes to other properties
  });
</script>
```

There's also a condition whereby the `product-category-load` is emitted before the related products island is initialized; in order to work around this, we'll store the information in the `#related-products-island-wrapper` element's `dataset` property:

```
<script>
  document.addEventListener('product-category-load',
    (event) => {
    const category = event.detail.category;
    const selectedProductId = event.detail.
      selectedProductId;
    Object.assign(
      document.querySelector('#related-products-island-
        wrapper').dataset,
      { category, selectedProductId }
    );
  });
</script>
```

We can then use that information as a mounting condition as well:

```
<is-land on:visible on:idle id="related-products-island-wrapper">
  <!-- no changes to template -->
  <script type="module">
    // no changes to the rest of the code
    const { selectedProductId, category } =
      document.querySelector(
      '#related-products-island-wrapper'
    ).dataset;

    mountRelatedProductsIsland(
      relatedProductsIslandContainer,
      category,
      selectedProductId,
      'data-*'
    );
  </script>
</is-land>
```

We render `from` to illustrate that both the `dataset`-based approach and the event-based approach both function in different scenarios.

If we load the page and scroll down to the related products (which are initially below outside the viewport), we'll see the following:

Related Products (from data-*)

Mens Cotton Jacket	Mens Casual Premium Slim Fit T-Shirts	Mens Casual Slim Fit
€55.99	€22.30	€15.99
4.7/5.0 (500)	4.1/5.0 (259)	2.1/5.0 (430)

Figure 6.14: The related products island with category information from data attributes

If we then reload the page, the scroll position will be such that the related products island is in view and initializes immediately, meaning the data comes from the custom event directly.

Related Products (from custom-event)

Mens Cotton Jacket	Mens Casual Premium Slim Fit T-Shirts	Mens Casual Slim Fit
€55.99	€22.30	€15.99
4.7/5.0 (500)	4.1/5.0 (259)	2.1/5.0 (430)

Figure 6.15: The related products island with category information from the custom event

We've now implemented the related products island with Preact and two approaches to reading the product category. Next, we'll see how to use bundling in conjunction with the islands architecture.

Scaling with a team – bundling islands

We can move the bulk of the code for a particular island to an external file and then use a tool such as esbuild to bundle it together. The following uses .jsx files for Preact, but a copy and paste of the existing files using htm would also work:

```
npx esbuild ./src/preact-cart-island.jsx --jsx-import-source=preact
--jsx=automatic --bundle --outdir=dist --format=esm --minify
npx esbuild ./src/vue-product-island.js --alias:vue=vue/dist/vue.esm-
bundler.js --bundle --outdir=dist --format=esm --minify
npx esbuild ./src/preact-related-products-island.jsx --jsx-import-
source=preact --jsx=automatic --bundle --outdir=dist --format=esm
--minify
```

The outputted files can then be used as follows:

```
<script type="module" src="./dist/
   preact-cart-island.js"></script>
<script type="module" src="./dist/
  vue-product-island.js"></script>
<script type="module">
  import { mountRelatedProductsIsland } from './dist/
    preact-related-products-island.js';
  // use mountRelatedProductsIsland
</script>
```

Each team can own one or more islands by providing a JavaScript bundle for them and/or a template (the template needs to be a server-side include).

Drawbacks

In the bundled use case, our two Preact islands don't share a Preact version, which means that this dependency will be loaded twice in the browser. This can be fixed with vendor bundles or module federation, as mentioned in the previous section. Also, note that it's not an issue for the initial version of the code where the scripts for the islands were in the page itself.

Challenges in an islands architecture mainly relate to component communication (as we've illustrated with the related products island) and the mechanism used to compose the templates and scripts in a unified page.

Summary

In this chapter, we've covered micro frontends, common approaches, and how the zones and islands architectures with Next.js and `is-land` allow us to build high-development velocity systems without compromising the user experience.

Micro frontends allow teams to have strong governance over different parts of a frontend ecosystem without compromising the user experience. Micro frontends allow more teams and their skills to be brought to bear effectively, which increases delivery velocity across the board. Common approaches include a container application with "server-side includes," build-time integration via shared packages, and runtime integrations (e.g., iframes, JavaScript, and Web Components).

The recommended Next.js "zones" approach allows different micro frontends to be mounted on different "base paths." The zones approach is a more flexible type of server-side includes; apps are "included" via a reverse-proxy and URLs. On a conceptual level, domain-specific applications that can deliver multiple pages and API routes are a great tool to leverage for larger teams.

Finally, we discussed the "islands" architecture implemented via the `is-land` package, which demonstrated a lightweight micro frontend approach with multiple JavaScript based libraries for different components. `is-land`'s ability to do partial hydration is a clear benefit to end users. Cross-island communication, a common challenge of the islands architecture, was addressed with an approach that includes `CustomEvent`'s and HTML data attributes.

Now that we've covered modern micro frontend approaches and the "zones" and "islands" architectures, we will look at patterns for performant asynchronous programming in JavaScript in the next chapter.

Part 3: Performance and Security Patterns

In this part, we will deep dive into performance and security patterns in JavaScript. You will learn how to optimize your asynchronous and event-driven JavaScript code for performance and security-sensitive contexts. In addition, you will learn about and implement asset-level optimizations, including lazy-loading and code-splitting JavaScript in a Next.js application, how to prioritize asset loading, and executing JavaScript off the main thread with Next.js and Partytown.

This part has the following chapters:

- *Chapter 7, Asynchronous Programming Performance Patterns*
- *Chapter 8, Event-Driven Programming Patterns*
- *Chapter 9, Maximizing Performance – Lazy Loading and Code Splitting*
- *Chapter 10, Asset Loading Strategies and Executing Code off the Main Thread*

7

Asynchronous Programming Performance Patterns

A key strength of JavaScript runtimes is the event loop, which couples "non-blocking input/output" within a single-threaded execution model. This means JavaScript is great for high-concurrency systems as long as they are not compute-bound systems (i.e., they're IO-bound).

With the asynchronous and non-blocking IO, JavaScript has strong built-ins to orchestrate requests. In this chapter, we'll cover the following topics:

- Sequential and parallel asynchronous operation patterns in JavaScript, both with Promises only and with async/await

- The cancellation and timeout of fetch requests with AbortController

- Advanced asynchronous operation patterns: throttling, debouncing, and batching

At the end of this chapter, you'll be able to spot and remedy situations where the asynchronous operation orchestration could be improved in JavaScript.

Technical requirements

You can find the code files for this chapter on GitHub at `https://github.com/PacktPublishing/Javascript-Design-Patterns`

Controlling sequential asynchronous operations with async/await and Promises

Promises were introduced in ES2015 (ES6), along with other modern data structures.

For those familiar with JavaScript prior to ES2015, asynchronous behavior was modeled with callback-based interfaces, for example, `request(url, (error, response) => { /* do work with response */ })`. The key issues that Promises resolved were the chaining of asynchronous requests and issues around managing parallel requests, which we'll cover in this section.

ES2016 included the initial specification for the async/await syntax. It built on top of the Promise object in order to write asynchronous code that didn't involve "Promise chains," where different Promises are processed using the `Promise().then` function. Promise functionality and async/await interoperate nicely. In fact, calling an async function returns a Promise.

We'll start by showing how to use Promises to manage sequential asynchronous operations. We'll use the Fetch API (which returns a Promise) to load `fakestoreapi.com/auth/login`. Given a username and password, and based on the output, we'll load all the relevant carts for that user. Subsequently, we'll load the relevant carts for that user using the `fakestoreapi.com/carts/user/{userId}` endpoint. This request flow is visualized in the following diagram.

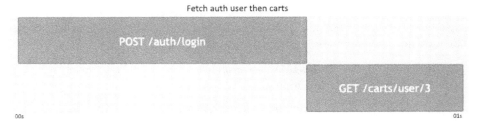

Figure 7.1: Sequence of /auth/login and /carts/user/{userId} requests

We'll start by sending a POST request to the `auth/login` endpoint. We add `.then((res) => res.json())`, which will wait for the initial `fetch()` output Promise to resolve to a "response" (hence the `res` name). We then call the `.json()` method on the response, which again is a Promise, which resolves to the JSON-decoded response body:

```
function fetchAuthUserThenCartsPromiseThen(username,
  password) {
  return fetch('https://fakestoreapi.com/auth/login', {
    method: 'POST',
    body: JSON.stringify({
      username,
      password,
    }),
    headers: {
      'Content-Type': 'application/json',
    },
  }).then((res) => res.json());
}
```

The Promise returned from res.json() can be accessed in another .then() callback, in which we parse the token field, which is a **JSON Web Token (JWT)**, using the jwt-decode package.

We extract the sub field from the decoded token. This is the "subject" claim, which tells us which user this token is about. In the case of the fakestoreapi token, userId is used as the "subject" claim. We can therefore use the sub claim as the user ID for which to load the carts in our following API call to https://fakestoreapi.com/carts/user/{userId}:

```
import jwt_decode from 'https://esm.sh/jwt-decode';

function fetchAuthUserThenCartsPromiseThen(username,
  password) {
  return // no change to the fetch() call
    .then((res) => res.json())
    .then((responseData) => {
      const parsedValues = jwt_decode(responseData.token);
      const userId = parsedValues.sub;
      return userId;
    })
    .then((userId) =>
      fetch(`https://fakestoreapi.com/carts/user/${userId}
        ?sort=desc`)
    )
    .then((res) => res.json());
}
```

This function can then be used as follows. Note that a password shouldn't be stored in the source of a production application (as it is in this example).

When we call the fetchAuthUserThenCartsPromiseThen function, it makes both the /auth/login call and then the /carts/user/{userId} call, which means we receive an array with the relevant carts for the requested user (note userId = 3, which is the correct ID for the kevinryan user).

Note that we're using async/await here to "flatten" the Promise output into userCartsDataPromiseThen, which we can assert on:

```
const username = 'kevinryan';
const password = 'kev02937@';

const userCartsDataPromiseThen = await
  fetchAuthUserThenCartsPromiseThen(
  username,
  password
);
```

```
assert.deepEqual(userCartsDataPromiseThen, [
  {
    __v: 0,
    date: '2020-01-01T00:00:00.000Z',
    id: 4,
    products: [
      {
        productId: 1,
        quantity: 4,
      },
    ],
    userId: 3,
  },
  {
    __v: 0,
    date: '2020-03-01T00:00:00.000Z',
    id: 5,
    products: [
      {
        productId: 7,
        quantity: 1,
      },
      {
        productId: 8,
        quantity: 1,
      },
    ],
    userId: 3,
  },
]);
```

As we've just seen in the code that calls fetchAuthUserThenCartsPromiseThen, the key benefit of async/await over Promise().then() chains is that the code is structured more similarly to synchronous code.

In synchronous code, the output of an operation can be, for example, assigned to a constant:

```
const output = syncGetAuthUserCarts();
console.log(output);
```

Whereas with Promise().then(), the output is available only in an additional .then callback:

```
promisifiedGetAuthUserCarts().then((output) => {
  console.log(output);
});
```

What `await` allows us to do is to structure the code as follows:

```
const output = await promisifiedGetAuthUserCarts();
console.log(output);
```

One way to think of it is that `await` can unfurl Promises. A Promise's "resolved value", usually only accessible in a `Promise().then()` callback is available directly.

For sequential operations, this is very useful, since it makes the code structured with a set of variable assignments per async operation.

The `await` operator is available at the top level of ECMAScript modules in modern runtime environments as part of the ES2022 specification.

However, in order to use `await` inside of a function, we need to mark the function as `async`. This usage of `await` in `async` functions has been available since ES2016.

Code editors and IDEs such as Visual Studio Code provide a refactor from chained `Promise().then()` calls to async/await. In our case, we can build a `fetchAuthUserThenCartsAsyncAwait` function as follows.

Instead of using `fetch().then(res => res.json())`, we'll first use `await fetch()` and then `await authResponse.json()`:

```
async function fetchAuthUserThenCartsAsyncAwait
  (username, password) {
  const authResponse = await fetch('https://fakestoreapi.com/auth/
login', {
    method: 'POST',
    body: JSON.stringify({
      username,
      password,
    }),
    headers: {
      'Content-Type': 'application/json',
    },
  });
  const authData = await authResponse.json();
}
```

We now have access to `authData`. We can decode `authData.token` as before using the `jwt-decode` package. This gives us access to the `sub` (subject) claim, which is the user ID:

```
Import jwt_decode from 'https://esm.sh/jwt-decode';

async function fetchAuthUserThenCartsAsyncAwait
  (username, password) {
```

```
    // no change to /auth/login API call code
    const parsedValues = jwt_decode(authData.token);
    const userId = parsedValues.sub;
}
```

Now that we have the relevant user ID, we can call the `/carts/user/{userId}` endpoint to load the user's carts:

```
async function fetchAuthUserThenCartsAsyncAwait
    (username, password) {
    // no change to /auth/login call or token parsing logic
    const userCartsResponse = await fetch(
        `https://fakestoreapi.com/carts/user/${userId}?sort=desc`
    );
    const userCartsResponseData = await userCartsResponse.
        json();
    return userCartsResponseData;
}
```

Given the same input data as the approach using `Promise().then()`, the loaded carts are the same. Note, again, that passwords and credentials should not be stored in source code files:

```
const username = 'kevinryan';
const password = 'kev02937@';

const userCartsDataAsyncAwait = await
fetchAuthUserThenCartsAsyncAwait(
    username,
    password
);
assert.deepEqual(userCartsDataAsyncAwait, userCartsDataPromiseThen);
```

One difference between the approaches is that with async/await, all the variables are defined in a single function scope, whereas the `Promise().then()` approach uses multiple function scopes (for each of the callbacks passed to `.then()`). With a single large function scope, variable names can't clash, which makes the code more verbose since, for example, each `response` object needs a qualifier to avoid variable name clashes, for example, `authResponse` and `userCartsResponse`.

The benefit of a single larger function scope is that all the outputs of previous API calls are available to subsequent ones without having to explicitly set them as values passed as a return in the callback passed to `.then()`.

Finally, a `fetch()`-specific example, is that since there are multiple Promises that require handling when doing a fetch and accessing the JSON response, the await approach can be a bit "noisier."

See the two following samples. First, with async/await, we assign a variable for the fetch `response` value:

```
const response = await fetch(url);
const data = await response.json();
```

Next, with `.then()`, we assign only a `data` variable and use an arrow function to handle the `.json()` unfurling:

```
const data = await fetch(url).then((response) => response.json());
```

As you see, our final example is a mix of `async/await` and `Promise().then()` so that the most "important" parts of the code are obvious. The specifics of how we extract the JSON output from `fetch` are not necessarily core to our logic so might be better expressed with `Promise().then()`.

In general, this slight difference in style wouldn't occur since parts of the code that are "less important," such as how we interact with the fetch API to process a request to JSON, tend to be abstracted – in this case, in an HTTP client of some kind. We would expect that the HTTP client could handle checking `response.ok` and accessing the response body as parsed JSON (using `response.json()`).

We've now seen how to implement sequential asynchronous operations using a Promise-only approach, an async/await-based approach, and finally, how both the async/await and Promise techniques can be used together to improve code readability and performance.

Parallel asynchronous operation patterns

A common source of bad performance is running operations sequentially that could be completed in parallel.

For example, a naive implementation of loading a cart and then the contained products would be as follows:

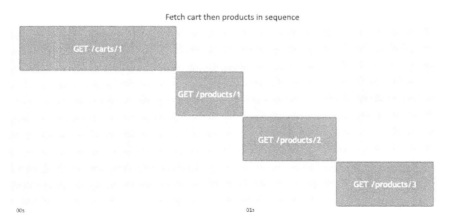

Figure 7.2: Load cart then each of the three products contained from fakestoreapi

In this case, the operation completion time is composed of the sum of the following:

- Request-response time for GET /carts/{cartId}
- Request-response time for GET /products/1
- Request-response time for GET /products/2
- Request-response time for GET /products/3

There is a requirement for the /products/{productId} calls to be done after the GET /carts/{cartId} call completes since that's where the product IDs are coming from. What isn't required is for each product call to wait for the previous one to complete; the calls only depend on data from the GET /carts/{cartId} call. This is an optimization opportunity. We can start all of the GET /products/{id} API calls together. We get the following sequence:

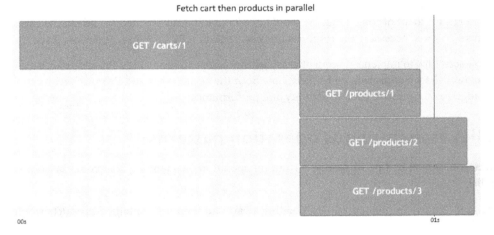

Figure 7.3: Load cart then each of the three products contained in parallel

In this case, the operation completion time is composed of the sum of the following:

- Request-response time for GET /carts/{cartId}
- The longest request-response time between GET /products/1, GET /products/2, and GET /products/3

This means we're saving the request-response time of two API calls at least.

JavaScript is especially well suited to these workloads since its concurrency model is based on an event loop. While JavaScript waits for an asynchronous operation to complete, it can complete other synchronous operations.

In layman's terms, triggering an asynchronous operation in JavaScript is "cheap and lightweight" compared to thread-based concurrency models that are common in popular programming languages such as Java and C++.

There are multiple constructs in JavaScript that allow us to convert an array of Promises into a Promise that resolves to an array. `Promise.all` is one such construct.

Implementing the scenario we described earlier, where we load a cart and then load the relevant product details, would look as follows with `Promise.all` and `Promise().then`.

First, we need to make the API call to load the cart and extract the JSON in the response body:

```
function fetchCartPromiseThen(cartId = '1') {
  return fetch(`https://fakestoreapi.com/carts/${cartId}`).
    then((res) =>
    res.json()
  );
}
```

We then need to set up the fetching of the right product URLs once the request to the `/carts/{cartId}` URL has loaded. The pattern used in order to execute our code after the fetch completes uses `.then()` on the returned promise:

```
function fetchCartPromiseThen(cartId = '1') {
    // no change to previous operations
    .then((cart) => {
      const productUrls = cart.products.map(
        (p) => `https://fakestoreapi.com/products/$
          {p.productId}`
      );
    })
}
```

Next, we'll use `Promise.all` to load all the product URLs with `fetch`. Since our goal is to return both the cart *and* the products, we'll return `{ cart }` as the first item in the array we're passing to `Promise.all()`. The rest of the array passed to `Promise.all` will be the Promises generated by calling `fetch().then((res) => res.json())` on each of the product URLs. In order to do this, we use the spread operation `(...)` on `...productUrls.map(/* mapping function */)` in the array:

```
function fetchCartPromiseThen(cartId = '1') {
    // no change to previous operations
    .then((cart) => {
      // no change to productUrls generation
      return Promise.all([
```

```
        { cart },
        ...productUrls.map((url) => fetch(url).then
            ((res) => res.json()))),
    ]);
    })
}
```

Finally, we're going to create an object with all the cart fields and a new products field based on the output of the /products/{id} fetches:

```
function fetchCartPromiseThen(cartId = '1') {
    // no change to previous operations
    .then(([prev, ...products]) => {
      return {
        ...prev,
        products,
      };
    });
}
```

We can test the output of the function by loading cart ID 1:

```
const cartDataFromPromiseThen = await fetchCartPromiseThen
    ('1');
```

The cart is as we expect – it returns three products:

```
assert.deepEqual(cartDataFromPromiseThen.cart, {
  __v: 0,
  date: '2020-03-02T00:00:00.000Z',
  id: 1,
  products: [
    {
      productId: 1,
      quantity: 4,
    },
    {
      productId: 2,
      quantity: 1,
    },
    {
      productId: 3,
      quantity: 6,
    },
  ],
```

```
  userId: 1,
});
```

The `products` field of our response contains the correct items in positions at indexes 0, 1, and 2:

```
assert.deepEqual(cartDataFromPromiseThen.products[0], {
  category: "men's clothing",
  description:
    'Your perfect pack for everyday use and walks in the
     forest. Stash your laptop (up to 15 inches) in the
     padded sleeve, your everyday',
  id: 1,
  image: 'https://fakestoreapi.com/img/
    81fPKd-2AYL._AC_SL1500_.jpg',
  price: 109.95,
  rating: {
    count: 120,
    rate: 3.9,
  },
  title: 'Fjallraven - Foldsack No. 1 Backpack, Fits 15
    Laptops',
});
assert.deepEqual(cartDataFromPromiseThen.
  products[1], {
  category: "men's clothing",
  description:
    'Slim-fitting style, contrast raglan long sleeve, three-button
henley placket, light weight & soft fabric for breathable and
comfortable wearing. And Solid stitched shirts with round neck made
for durability and a great fit for casual fashion wear and diehard
baseball fans. The Henley style round neckline includes a three-button
placket.',
  id: 2,
  image:
    'https://fakestoreapi.com/img/
       71-3HjGNDUL._AC_SY879._SX._UX._SY._UY_.jpg',
  price: 22.3,
  rating: {
    count: 259,
    rate: 4.1,
  },
  title: 'Mens Casual Premium Slim Fit T-Shirts ',
});
assert.deepEqual(cartDataFromPromiseThen.products[2], {
  category: "men's clothing",
```

```
description:
    'great outerwear jackets for Spring/Autumn/Winter, suitable for
many occasions, such as working, hiking, camping, mountain/rock
climbing, cycling, traveling or other outdoors. Good gift choice for
you or your family member. A warm hearted love to Father, husband or
son in this thanksgiving or Christmas Day.',
    id: 3,
    image: 'https://fakestoreapi.com/img/
      71li-ujtlUL._AC_UX679_.jpg',
    price: 55.99,
    rating: {
      count: 500,
      rate: 4.7,
    },
    title: 'Mens Cotton Jacket',
});
```

We've now seen how to leverage `Promise.all` to run multiple promises in parallel and handle their output with one handler.

You'll have noticed the "trick" we did by passing the `{ cart }` object in `Promise.all` and then extracting the first item of the resolved array as the previous response. This is a limitation of `Promise().then()` chaining, as mentioned in the *Controlling sequential asynchronous operations with async/await and Promises* section. Each function argument to `.then()` gets its own scope:

```
Promise.resolve({ id: 1 })
  .then((cart) => {
    const productUrls = [];
    return Promise.all([{ cart }, ...productUrls.map(()
      => {})]);
  })
  .then(([prev, ...products]) => {});
```

An alternative way to write this is to store the cart in the function scope:

```
function fetchCartFunctionVariable() {
  let loadedCart = null;
  return Promise.resolve({ id: 1 })
    .then((cart) => {
      loadedCart = cart;
      const productUrls = [];
      return Promise.all(productUrls.map(() => {}));
    })
    .then((products) => ({
      cart: loadedCart,
```

```
      products,
    }));
}
```

This works as expected. We've obviously removed the actual cart and product fetching logic from the API, but the cart that `{ id: 1 }` relates to, which we resolved in the initial `Promise.resolve()` function call, is cached through the `.then()` calls:

```
assert.deepEqual(await fetchCartFunctionVariable(), {
  cart: { id: 1 },
  products: [],
});
```

Another way to improve our implementation without resorting to function-scoped variables, which can be hard to keep track of, is to convert it to use async/await.

Our logic would be as follows. We start by loading the cart and converting the JSON response body:

```
async function fetchCartAsyncAwait(cartId = '1') {
  const cart = await fetch(`https://fakestoreapi.com/
carts/${cartId}`).then(
    (res) => res.json()
  );
}
```

Once the cart is loaded, we proceed to fetch the relevant products by generating URLs based on the `cart.products` contents (mainly the `productId` field). We also fetch these URLs using `Promise.all`:

```
async function fetchCartAsyncAwait(cartId = '1') {
  // no change to cart fetching
  const productUrls = cart.products.map(
    (p) => `https://fakestoreapi.com/products/${p.productId}`
  );
  const products = await Promise.all(
    productUrls.map((url) => fetch(url).then((res)
      => res.json()))
  );
}
```

Finally, we can return the cart and the loaded products:

```
async function fetchCartAsyncAwait(cartId = '1') {
  // no changes to cart or products fetching
  return {
    cart,
    products,
```

```
  };
}
```

The implementation is equivalent to our previous, strict `Promise().then()`-based one, as the following checks attest:

```
const cartDataFromAsyncAwait = await fetchCartAsyncAwait
  ('1');

assert.deepEqual(cartDataFromPromiseThen.cart,
  cartDataFromAsyncAwait.cart);
assert.deepEqual(
  cartDataFromPromiseThen.products,
  cartDataFromAsyncAwait.products
);
```

The benefit of using async/await in this case was, again, an increase in readability. The syntax gets less in the way than chained `.then()` calls and we don't have to resort to either returning the first response as an item in `Promise.all([{ cart }])` or adding a function-scoped variable that we store the cart in.

We've now seen how to leverage `Promise.all` to complete asynchronous operations in parallel both with a `Promise().then()`-exclusive approach and with judicious refactors to `async/await` to simplify the code.

Next, we'll see how we can cancel and time out requests with `AbortController` in JavaScript.

Asynchronous cancellation and timeouts with AbortController

Another source of bad performance in applications in general is doing work that's not necessary. In the context of a JavaScript web application, one of the types of "work" that can be unnecessary (and therefore a drain on performance) is having HTTP requests that aren't required any more. For example, in a photo gallery system or any paginated system, when moving across photos, the request for the previous photo might not have completed before the next one is started. In this case, the previous request data is not necessary any more, as we're essentially on a completely different page.

In these instances, cancelling the request might be useful.

`AbortController` is a Web/DOM API that allows us to abort web requests. It's created using its constructor, `new AbortController`, and controlling a request (to potentially cancel it) is done with the `AbortController().signal` value, which is an `AbortSignal` object.

We instantiate the controller using the new `AbortController()` constructor call. If we want to make a `fetch` call cancellable, we pass `abortController.signal` as the `signal` option:

```
function fetchWithCancel(url) {
  const abortController = new AbortController();

  const response = fetch(url, { signal:
    abortController.signal }).then((res) =>
    res.json()
  );
  return {
    response,
  };
}
```

If we want to cancel the `fetch` request, we can then call `abortController.cancel`. We'll add this as a `cancel` function on the `fetchWithCancel` returned output:

```
function fetchWithCancel(url) {
  // no changes to contents
  return {
    // no changes to other keys in the object
    cancel: () => abortController.abort(),
  };
}
```

Finally, we need to ensure that when we see `AbortError`, we handle it. In this case, we'll handle it with a `Promise().catch` handler, which, on seeing an `AbortError`, will return 'Aborted', and re-throw the error otherwise.

An `AbortError` error instance has a name property equal to `'AbortError'`, but also a message such as `DOMException [AbortError]: This operation was aborted`, along with its stack trace:

```
function fetchWithCancel(url) {
  // no change to abortController initiationisalition
  const response = fetch(url, { signal: abortController.signal })
    .then((res) => res.json())
    .catch((err) => {
      if (err.name === 'AbortError') return 'Aborted';
      throw err;
    });
  // no change to return value
}
```

Given two API calls to fakestoreapi, /products/1, and /products/2, we can cancel one of them without affecting the other request as follows, by calling fetchWithCancel with both URLs and storing the output in two variables. Note that we're not using await yet.

We can then cancel the fetch for /products/1 by using the .cancel() function we built earlier:

```
const fetchProduct1 = fetchWithCancel
  ('https://fakestoreapi.com/products/1');
const fetchProduct2 = fetchWithCancel('https://fakestoreapi.com/
products/2');

fetchProduct1.cancel();
```

The outcome of this is that when we await fetchProduct1.response and fetchProduct2.response, the output for fetchProduct1.response is 'Aborted', which means an AbortError instance was handled in fetchWithCancel (i.e., our cancellation succeeded).

The output for fetchProduct2.response is the product object:

```
assert.deepEqual(await fetchProduct1.response, 'Aborted');
assert.deepEqual(await fetchProduct2.response, {
  category: "men's clothing",
  description:
    'Slim-fitting style, contrast raglan long sleeve, three-button
henley placket, light weight & soft fabric for breathable and
comfortable wearing. And Solid stitched shirts with round neck made
for durability and a great fit for casual fashion wear and diehard
baseball fans. The Henley style round neckline includes a three-button
placket.',
  id: 2,
  image:
    'https://fakestoreapi.com/img
      /71-3HjGNDUL._AC_SY879._SX._UX._SY._UY_.jpg',
  price: 22.3,
  rating: {
    count: 259,
    rate: 4.1,
  },
  title: 'Mens Casual Premium Slim Fit T-Shirts ',
});
```

Manually cancelling a request is useful, but a more widespread use case is to time a request out when it takes more than a certain amount of time. This is useful to ensure a responsive user experience for customers. Different situations call for longer or shorter timeout delays.

We can implement a fetchWithTimeout function using fetch, AbortController, and setTimeout.

Our function takes a URL and an optional timeout, which we'll default to 500 (for 500 ms). Similar to our manual cancellation scenario (see fetchWithCancel), we'll create an abortController object and pass its signal property as an option to fetch:

```
async function fetchWithTimeout(url, timeout = 500) {
  const abortController = new AbortController();
  return fetch(url, { signal: abortController.signal });
}
```

In order to cancel the fetch after a certain amount of time, we'll use setTimeout. The setTimeout handler will simply call abortController.abort() and we'll set the timeout delay to our timeout variable:

```
async function fetchWithTimeout(url, timeout = 500) {
  // no change to abortController
  setTimeout(() => {
    abortController.abort();
  }, timeout);
  // no change to fetch call or return
}
```

When the request takes less time than the fetch request takes to complete, we receive the response data:

```
const timedoutFetchShouldSucceedData = await fetchWithTimeout(
  'https://fakestoreapi.com/products/1',500
)
  .then((res) => res.json())
  .catch((error) => {
    if (error.name === 'AbortError') {
      return 'Aborted';
    }
    throw error;
  });

console.assert(
  timedoutFetchShouldSucceedData.id === 1,
  'fetchWithTimeout with 500ms timeout should have
    succeeded'
);
```

When a fetch request takes longer than the configured timeout, we receive an AbortError instance:

```
const timedoutFetchShouldAbort = await fetchWithTimeout(
  'https://fakestoreapi.com/products/1',10
)
```

```
  .then((res) => res.json())
  .catch((error) => {
    if (error.name === 'AbortError') {
      return 'Aborted';
    }
    throw error;
  });

console.assert(
  timedoutFetchShouldAbort === 'Aborted',
  'fetchWithTimeout with 10ms timeout should have
    aborted but did not'
);
```

We've now seen how to use `AbortController` to control `fetch` cancellation manually and how to use it to create a "fetch with timeout" utility. We can use `AbortController` to cancel operations that aren't required any more, thereby reducing network usage.

Next, we'll look at further patterns that can help optimize situations with high volumes of requests.

Throttling, debouncing, and batching asynchronous operations

Throttling is an operation in which requests are dropped until a certain time is reached. For example, for a 10 ms throttle timeout, once a request is made, no request in the next 10 ms will be sent. If multiple requests are made between 0 ms and 10 ms, only the last request will be sent after the 10 ms timeout expires.

In JavaScript, such a throttle function can be implemented as follows.

A higher-order function, `throttle` takes in an `fn` parameter and returns an executable function with the same input signature as the `fn` parameter.

When the "throttled" `fn` function is called, we set `isThrottled = true` in order to be able to discard calls between the first call and a configured timeout:

```
function throttle(fn, timeout) {
  let isThrottled = false;
  return (...args) => {
    isThrottled = true;
    return fn(...args);
  };
}
```

We now need to ensure `fn` is not called while `isThrottled` is true. We achieve this by returning early from our returned "throttled" `fn` function.

We save the arguments with which the "throttled" `fn` function was called so that they can be used when the timeout expires:

```
function throttle(fn, timeout) {
  // no change to existing variable definitions
  let lastCallArgs = null;
  return (...args) => {
    if (isThrottled) {
      lastCallArgs = args;
      return;
    }

    // no change to "initial call" case
  };
}
```

Finally, we configure `setTimeout` to trigger a reset of the throttled state and execute the last function call:

```
function throttle(fn, timeout) {
  // no change to existing variable definitions
  return (...args) => {
    // no change to short-circuit logic
    setTimeout(() => {
      isThrottled = false;
      return fn(...lastCallArgs);
    }, timeout);
    // no change to "initial call" case
  };
}
```

A simple example of this in use is the following scenario, where many messages could be sent in a given time. Instead, we want to throttle to 1 message every 1 ms interval.

Our `storeMessage` function is as follows:

```
let messages = [];
const storeMessage = (message) => {
  messages.push(message);
};
```

We can generate a `throttledStoreMessage` function with a 1 ms timeout as follows.

When called ten times synchronously and subsequently waiting for timers to complete, only the first (`'throttle-1'`) and last (`'throttle-10'`) calls are recorded:

```
const throttledStoreMessage = throttle(storeMessage, 1);
throttledStoreMessage('throttle-1');
throttledStoreMessage('throttle-2');
throttledStoreMessage('throttle-3');
throttledStoreMessage('throttle-4');
throttledStoreMessage('throttle-5');
throttledStoreMessage('throttle-6');
throttledStoreMessage('throttle-7');
throttledStoreMessage('throttle-8');
throttledStoreMessage('throttle-9');
throttledStoreMessage('throttle-10');

await timeout();
assert.deepEqual(messages, ['throttle-1', 'throttle-10']);

function timeout(ms = 0) {
  return new Promise((r) => setTimeout(r, ms));
}
```

If we reset the messages and wait for the timers to complete after our call with `'throttle-5'`, we finish with `['throttle-1', 'throttle-5', 'throttle-6']`, that is, the first call, and the calls before and after the timers are cleared.

If we clear the timers one more time after completing all our calls, `'throttle-10'` is also present in our messages list, meaning that call completed:

```
messages = [];
throttledStoreMessage('throttle-1');
throttledStoreMessage('throttle-2');
throttledStoreMessage('throttle-3');
throttledStoreMessage('throttle-4');
throttledStoreMessage('throttle-5');
await timeout();
throttledStoreMessage('throttle-6');
throttledStoreMessage('throttle-7');
throttledStoreMessage('throttle-8');
throttledStoreMessage('throttle-9');
throttledStoreMessage('throttle-10');

assert.deepEqual(messages, ['throttle-1', 'throttle-5',
'throttle-6']);
await timeout();
```

```
assert.deepEqual(messages, [
  'throttle-1',
  'throttle-5',
  'throttle-6',
  'throttle-10',
]);
```

We've now seen how to throttle a function. We can now look at debouncing.

A `debounce` function in JavaScript takes an `fn` parameter, which is a function. The goal is that the debounced `fn` function should discard all calls except the last call before it's not called for a `timeout` period.

In order to do this, we should "delay" the function call until after a timeout completes. We save the `timeoutId` reference in order to cancel the call if the debounced `fn` function is called again. We use `setTimeout` and forward the arguments with which the debounced `fn` function was called:

```
function debounce(fn, timeout) {
  let timeoutId;
  return (...args) => {
    timeoutId = setTimeout(() => {
      fn(...args);
    }, timeout);
  };
}
```

With the current state of the `debounce` function, there would still be as many calls to `fn` as there are to the debounced `fn` function; they would just be queued for delayed execution based on the timeout. To avoid this, we can cancel the previous call timeout by using `clearTimeout(timeoutId)`:

```
function debounce(fn, timeout) {
  // no change to variable declarations
  return (...args) => {
    clearTimeout(timeoutId);
    // no change to setTimeout logic
  };
}
```

With these changes in place, if we create a `debouncedStoredMessage` function with a 1 ms timeout and call it 10 times, it will not execute until we wait for the timer to complete:

```
messages = [];
const debouncedStoredMessage = debounce(storeMessage, 1);
debouncedStoredMessage('debounce-1');
debouncedStoredMessage('debounce-2');
```

```
debouncedStoredMessage('debounce-3');
debouncedStoredMessage('debounce-4');
debouncedStoredMessage('debounce-5');
debouncedStoredMessage('debounce-6');
debouncedStoredMessage('debounce-7');
debouncedStoredMessage('debounce-8');
debouncedStoredMessage('debounce-9');
debouncedStoredMessage('debounce-10');

assert.deepEqual(messages, []);
await timeout();
assert.deepEqual(messages, ['debounce-10']);
```

We can further showcase this by waiting for timers to complete after the fifth call. In that case, the fifth call will trigger and, given another timeout window clears, the tenth call will also trigger:

```
messages = [];
debouncedStoredMessage('debounce-1');
debouncedStoredMessage('debounce-2');
debouncedStoredMessage('debounce-3');
debouncedStoredMessage('debounce-4');
debouncedStoredMessage('debounce-5');
await timeout();
debouncedStoredMessage('debounce-6');
debouncedStoredMessage('debounce-7');
debouncedStoredMessage('debounce-8');
debouncedStoredMessage('debounce-9');
debouncedStoredMessage('debounce-10');

assert.deepEqual(messages, ['debounce-5']);
await timeout();
assert.deepEqual(messages, ['debounce-5', 'debounce-10']);
```

We've now seen how to throttle and debounce functions, which allows us to ensure operations don't trigger more than necessary.

In a scenario where we have a "search as you type" or "suggest as you type" input (sometimes referred to as a "typeahead"), which needs to make API requests to get search results or suggestions, it usually makes sense to use either debounce, to wait for the user to stop typing before making a request, or to throttle the requests so that an API request is made every window instead of every keystroke.

This can also be coupled with other heuristics to avoid overwhelming the API server with unnecessary requests. For example, it's usual to avoid sending requests until a few characters have been typed since the search request is too broad with only 1 or 2 characters.

We've seen how to protect an API by reducing the number of requests using throttling or debouncing. In the *Parallel asynchronous operation patterns* section, we used `Promise.all` to send requests in parallel. This can be another scenario where the target of our asynchronous operations can get overwhelmed. To avoid an overload scenario, it can be useful to batch our requests.

"Batching" is a way to limit concurrency, for example, instead of sending 20 requests at the same time (in parallel), we want to send 5 at a time.

A `batch` function takes an array and a batch size and returns an array of arrays. The nested arrays have a maximum length of "batch size."

We start by calculating how many `batchItem` list items we'll need in our `batches` array. In order to do this, we divide the input array length by the batch size and apply the `ceil` function to the value. In other words, we round up `inputLength` divided by `batchSize` to the next largest integer value.

We can then generate our `batches` array with the right size (`batchCount`, as computed):

```
function batch(inputArray, batchSize) {
  const batchCount = Math.ceil(inputArray.length /
    batchSize);
  const batches = Array.from({ length: batchCount });
}
```

We then go through each of the batches using `Array.prototype.map()`. The items in `batches` are initially undefined, but we use the index of the item (which we'll call `batchNumber`). For each item in `batches`, we take the items from `batchNumber * batchSize` to `(batchNumber + 1) * batchSize` and they constitute the contents of our `batches[batchNumber]` array item:

```
function batch(inputArray, batchSize) {
  // no change to existing size computations
  return batches.map((_, batchNumber) => {
    return inputArray.slice(
      batchNumber * batchSize,
      (batchNumber + 1) * batchSize
    );
  });
}
```

You'll note that we're generating the array with `Array.from` *and then* populating it using `Array.prototype.map()`, however, `Array.from()` supports a second parameter, which is a mapping function. Our code could therefore be as follows:

```
function batch(inputArray, batchSize) {
  const batchCount = Math.ceil(inputArray.length /
    batchSize);
  return Array.from({ length: batchCount }, (_,
```

```
    batchNumber) => {
    return inputArray.slice(
      batchNumber * batchSize,
      (batchNumber + 1) * batchSize
    );
  });
}
```

In any case, our `batch` function work for any array, for example, a 10-element array can be batched into chunks of 4 or 3 correctly by our function:

```
assert.deepEqual(batch([1, 2, 3, 4, 5, 6, 7, 8, 9, 10, 11], 4), [
  [1, 2, 3, 4],
  [5, 6, 7, 8],
  [9, 10, 11],
]);
assert.deepEqual(batch([1, 2, 3, 4, 5, 6, 7, 8, 9, 10, 11], 3), [
  [1, 2, 3],
  [4, 5, 6],
  [7, 8, 9],
  [10, 11],
]);
```

The preceding example showcased a synchronous example. For our use case – improving the performance of asynchronous operations, we need to handle Promises. The good news is that Promises can be stored in an array just as well:

```
const numberResolverBatches = batch(
  [Promise.resolve(1), Promise.resolve(2), Promise.resolve(3)],
  2
);

console.assert(numberResolverBatches.length === 2);
console.assert(numberResolverBatches[0].length === 2);
console.assert(numberResolverBatches[1].length === 1);
```

To get the batched output of the Promises, however, we need to write a function that awaits all the Promises in each batch to resolve them sequentially.

This can be achieved by using a `for ... of` loop and `Promise.all`, as follows. We flatten out the resolved values:

```
async function resolveBatches(batchedPromises) {
  const flattenedBatchOutput = [];
  for (const batch of batchedPromises) {
```

```
    const resolved = await Promise.all(batch);
    flattenedBatchOutput.push(...resolved);
  }
  return flattenedBatchOutput;
}

const batchOutput = await resolveBatches(numberResolverBatches);
assert.deepEqual(batchOutput, [1, 2, 3]);
```

In our example, the `Promise.resolve()` calls with 1, 2, and 3 can indeed be batched and resolved.

We've now seen how to build and use throttling, debouncing, and batching to improve the performance of our asynchronous operations in JavaScript.

Summary

In this chapter, we've covered asynchronous operation orchestration patterns with Promises and async/await to manage sequential and parallel operations. We also covered advanced patterns such as request cancellation, implementing timeouts, the difference between throttling and debouncing, and finally, how to use batching in an asynchronous operation context.

In order to manage sequential asynchronous operations, we can use a Promise-based approach with `Promise().then()`, async/await, or mix both approaches. This helps keep our code simple to reason about. For parallel execution, we can leverage `Promise.all()` with `Promise.then()` or async/await. We also have multiple approaches to maintaining response data across asynchronous operations.

We can leverage `AbortController` to cancel requests. We implemented a timeout for the `fetch` response time using `AbortController` and `setTimeout`. Stopping in-flight requests is a useful cleanup step that can improve performance by reducing unnecessary load on our API origin.

Finally, the advanced asynchronous programming patterns allow fewer requests to happen via throttling and debouncing. We can also control the concurrency of our parallel requests using batching and resolving the batches. Again, these approaches can reduce unnecessary network traffic and load on the API servers.

Now that we've covered asynchronous programming performance patterns, with Promise, async/await, and advanced patterns, we can look at patterns for event-driven programming in JavaScript.

8

Event-Driven Programming Patterns

Event-driven programming in JavaScript is very widespread and is the only way to handle certain scenarios. Maintaining performance and security around event listeners is of paramount importance. Mismanaged event listeners have been a historical source of bugs and critical performance issues; we'll address this via the event delegation pattern. Secure messaging between frames and contexts has always been crucial in the context of payments. More recently, new primitives are being added to the web platform and JavaScript that exposes an event/messaging interface for maintaining isolation between contexts.

In this chapter, we'll cover the following topics:

- Implementing event delegation
- Using the `postMessage` interface to communicate across contexts with an example of a payment iframe
- Common event listener antipatterns and how to remediate them

At the end of this chapter, you'll have learned how to use advanced event-driven programming concepts in JavaScript to keep your code performant and secure.

Technical requirements

You can find the code files for this chapter on GitHub at `https://github.com/PacktPublishing/Javascript-Design-Patterns`

Optimizing event listeners through event delegation

Event delegation is a common event listener pattern used to go from "many elements, many event listeners" to a "many elements, single event listener." At its core, event delegation attaches one event listener to the page's `Document`, and inside that listener, it checks what the `target` of the event is in order to figure out how the event should be handled.

Event delegation means fewer listeners are attached. There's only one per root node; if we're doing event delegation at the document level, that means one listener. Another benefit is that DOM nodes can be attached and removed without worrying about adding or removing the relevant event listeners.

The following sequence diagram details an implementation of listening to clicks on two buttons.

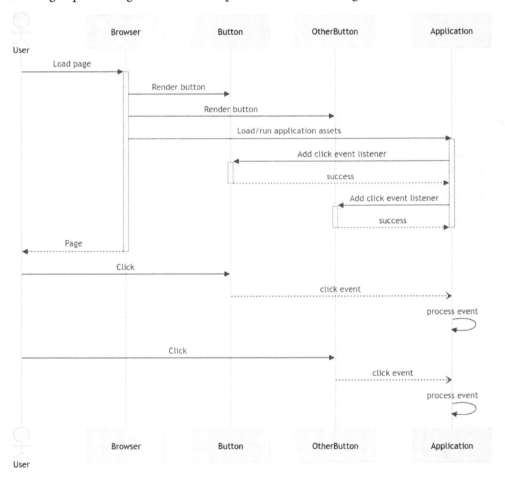

Figure 8.1: Event handling without event delegation

Event handling without event delegation can be contrasted with the event delegation sequence, which instead of attaching one handler per event/element, attaches one and computes the relevant action in the single listener.

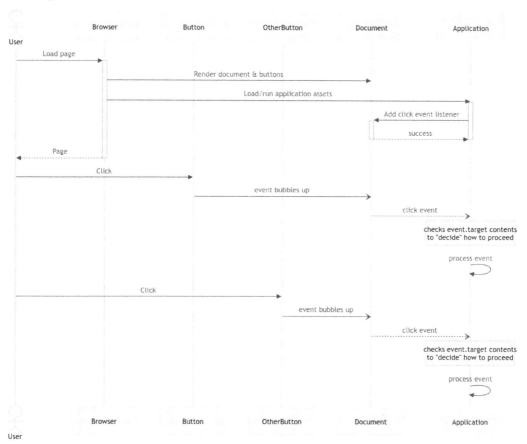

Figure 8.2: Event handling with event delegation

We'll implement simple event delegation for an email subscription form that is submitted via client-side JavaScript with `fetch`. To begin with, we'll start laying out a form. We have a form with the `data-newsletter-form` attribute, which we'll pick up in JavaScript, a heading, a label, an email input, and a submit button:

```
<form data-newsletter-form>
  <h3>Subscribe to the newsletter!</h3>
  <div>
    <label for="email">Email</label>
    <input
```

```
        id="email"
        type="email"
        name="email"
        placeholder="test@example.com"
      />
    </div>
    <button type="submit">Submit</button>
  </form>
```

To start on the event part of event delegation, we add a click listener to the document. This listener switches based on event.target.tagName; tagName takes uppercase values such as P, BUTTON, and DIV. In order to make the code easier to follow, we'll switch on the lowercase version of tagName:

```
<script>
  document.addEventListener('click', (event) => {
    switch (event.target.tagName?.toLowerCase()) {
    }
  });
</script>
```

When we detect a click on a button element, we then check whether the event.target is inside a form, whether event.target is of the submit type, and whether the form that contains the event target element includes newsletterForm in its dataset, in other words whether it has data-newsletter-form. In this case, we call event.preventDefault. We'll be handling the form submission using JavaScript.

We provide some feedback to the user that the form is submitting by changing the contents of the event target button (to Submitting) and we set the disabled attribute so that the form can't be submitted again until our handler execution completes:

```
<script>
  document.addEventListener('click', (event) => {
    switch (event.target.tagName?.toLowerCase()) {
      case 'button': {
        const form = event.target.closest('form');
        if (
          form &&
          event.target.type === 'submit' &&
          'newsletterForm' in form.dataset
        ) {
          event.preventDefault();

          const formValues = new FormData(form);
```

```
           event.target.innerText = 'Submitting';
           event.target.setAttribute('disabled',
             'disabled');

           const email = formValues.get('email');
           return;
         }
       }
     }
   });
</script>
```

When we click the **Submit** button, it now gets disabled, and its content is set to **Submitting**.

Figure 8.3: When the Submit button is clicked, it is disabled and the text changes to Submitting

We'll now work on submitting the newsletter form. In order to do this, we need a `fetch`-based function that will `POST` the given `email` parameter to `jsonplaceholder.typicode.com/users`. We then await the `fetch` promise and extract the JSON response using `res.json()`:

```
<script>
  async function submitNewsletterSubscription(email) {
    const res = await fetch
      ('https://jsonplaceholder.typicode.com/users', {
      method: 'POST',
      body: JSON.stringify({
        email,
      }),
      headers: {
        'Content-type': 'application/json; charset=UTF-8',
      },
    });
    return res.json();
  }
  // no change to the document "click" event listener
</script>
```

We'll now extend the `button type=submit` handler to call `submitNewsletterSubscription`. The `email` value comes from `formValues.get('email')` (the email field of the form). when `submitNewsletterSubscription` completes successfully (i.e.. the Promise resolves), we reset the `submit` button to have the text **Submit** and to be enabled (by removing the `disabled` attribute):

```
<script>
  document.addEventListener('click', (event) => {
    switch (event.target.tagName?.toLowerCase()) {
      case 'button': {
        const form = event.target.closest('form');
        if (
          form &&
          event.target.type === 'submit' &&
          'newsletterForm' in form.dataset
        ) {
          // no change to existing logic
          const email = formValues.get('email');

          submitNewsletterSubscription(email).then((result) => {
            event.target.innerText = 'Submit';
            event.target.removeAttribute('disabled');
          });
        }
        return;
      }
    }
  });
</script>
```

To highlight the requests/responses, we'll add a `storeLogEvent` function and an API request/response log to our page:

```
<div style="height: 300px; overflow: scroll">
  <h3>API Request/Response Log</h3>
  <pre><code></code></pre>
</div>
<script>
  // no change to other functionality
  function storeLogEvent(value) {
    $requestLog = document.querySelector('pre code');
    $requestLog.innerText += value;
    $logParent = $requestLog.closest('div');
    $logParent.scrollTo({ top: $logParent.scrollTopMax,
      behavior: 'smooth' });
```

```
    }
</script>
```

We can then use `storeLogEvent` before and after calling `submitNewsletterSubscription`:

```
// -> inside the listener
// -> switch
// -> case 'button'
// -> if (form in ancestors && button type === submit &&
   form has data-newsletter-form)
const email = formValues.get('email');
storeLogEvent(`Request: ${email}`);
submitNewsletterSubscription(email).then((result) => {
  storeLogEvent(`\nResponse: ${JSON.stringify(result,
    null, 2)}\n\n`);

  event.target.innerText = 'Submit';
  event.target.removeAttribute('disabled');
});
```

Now, when we click **Submit** with an email address in the input, it gets POST-ed to `jsonplaceholder` and we get a response back, as we can see in the following screenshot:

Figure 8.4: When we enter an email and click Submit, we the API response is displayed

To showcase the benefit of event delegation in a situation where DOM elements can get dynamically added, we'll create an **Add a form** button that will append an additional newsletter form to the document.

First, we add a `button` with `data-add-form`, we'll use the data attribute to detect and handle clicks on the `button`.

```
<button data-add-form>Add a form!</button>
```

Next, we'll handle clicks on elements with our `data-add-form` attribute by adding `if ('addForm' in event.target.dataset)`. For now, we'll return early to prevent any further handling code from executing:

```
<script>
  document.addEventListener('click', (event) => {
    switch (event.target.tagName?.toLowerCase()) {
      case 'button': {
        if ('addForm' in event.target.dataset) {
          return;
        }
        // no change to newsletter form handling
      }
    }
  });
</script>
```

We want to implement the add form functionality, and we want to find a `data-newsletter-form` element and clone it using `.cloneNode(true)`.

We'll append a random number inside the heading so we can identify when new forms are added and reset the email input. Finally, we append the new node to the `document.body` element using `.appendChild`:

```
<script>
  document.addEventListener('click', (event) => {
    switch (event.target.tagName?.toLowerCase()) {
      case 'button': {
        if ('addForm' in event.target.dataset) {
          const $newsletterFormTemplate = document.
            querySelector(
            '[data-newsletter-form]',
          );
          const newForm = $newsletterFormTemplate.
            cloneNode(true);
          newForm.querySelector('h3').innerText += `
            (${Math.floor(
```

```
        Math.random() * 100,
    )})`;
    newForm.querySelector('[name=email]').value = '';
    document.body.appendChild(newForm);
    return;
  }
  // no change to newsletter form handling
}
  }
  });
</script>
```

With no changes to the handling of the newsletter form submission, the cloned form functions just as the initial one does.

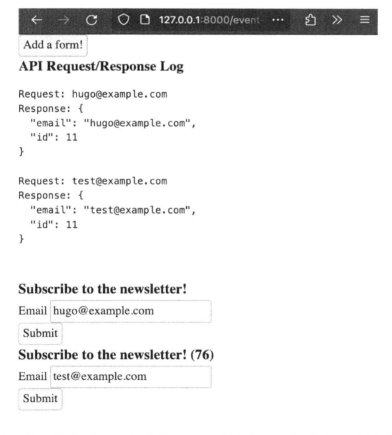

Figure 8.5: The effect of submit on multiple forms, one of which was added with an Add a form! button

We've now seen how to implement event delegation to prevent having to add event listeners manually to dynamically added DOM nodes. Next, we'll look at patterns that use the `postMessage` interface between iframes.

Patterns for secure frame/native WebView bridge messaging

Gaining a deep understanding of messaging patterns with `postMessage` in JavaScript is crucial for working in a variety of contexts. `postMessage` is defined on the following Web API objects: `Window`, `MessagePort`, `Worker`, `Client`, `ServiceWorker`, and `BroadcastChannel`.

In other words, `postMessage`-based messaging is useful for document-to-iframe, iframe-to-iframe, document-to-worker, and service worker-to-document communication and that's only the Web APIs. Due to how widespread the `postMessage` API is, it's also adopted in non-standard APIs for handling multiple JavaScript contexts. For example, web extensions for Chrome and Firefox contain multiple JavaScript contexts: the devtools panel, proxy, backend, and background script. The `postMessage` API is also used for Android and iOS communication between the native code and WebViews.

The scenario that we'll go through is about iframes and how they communicate. A common e-commerce use-case is integrating a third-party payment service provider's hosted card capture form into their e-commerce website. By using a payment service provider and not knowing the customer's card payment details, the e-commerce vendor can meet **Payment Card Industry Data Security Standard (PCI DSS)** compliance more easily.

The container or parent document will be a checkout form, inside of which we'll iframe a hosted card capture document. The two documents will communicate with `postMessage`. The container document will not read the card details in cleartext. Instead, it will receive a public-key encrypted payload (which can only be decrypted via the paired private key).

Without being careful, it's possible for `iframe` initialization to cause race conditions. To work around this, we'll implement the following initialization scheme.

Initially, we'll load a container document with an `iframe` that has no `src`. Only after we've added important event listeners to the `iframe` element, will we add `src`. This means that the `iframe` can't load before our listeners are attached.

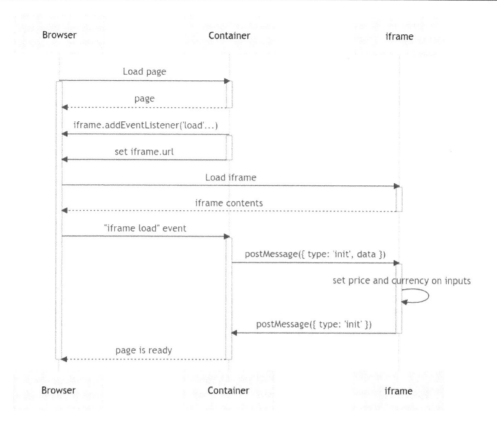

Figure 8.6: Sequence diagram of initialization messaging

We need two files, one at `frame-parent.html` (which will be our application shell) and one at `frame-content.html` (which will represent our iframe's contents).

Some payment service provider integrations won't require a fully custom `iframe` (sometimes, a JavaScript SDK is provided that helps manage the `iframe` part of it), but the important thing is that the `iframe` is loaded from an origin (server) that is owned by the payment service provider. We won't be able to represent this since we're working locally.

Our `frame-parent.html` HTML looks as follows: a few headings, a `form`, an `input type=email`, an `iframe`, and a submit button. Note that the `iframe` element doesn't have a `src` attribute. We'll add that via JavaScript to prevent race conditions:

```
<form>
  <h2>Checkout form</h2>
  <div>
    <span>Price: <span>200</span>€</span>
```

```
      </div>
      <div>
        <h3>Customer Details</h3>
        <div>
          <label for="email">Email</label>
          <input id="email" type="email" name="email"
            required="required" />
        </div>
      </div>
      <div>
        <iframe id="payment-capture" width="100%"
          height="300px"></iframe>
      </div>
      <div>
        <button type="submit">Pay</button>
      </div>
    </form>
```

To prevent race conditions when loading the iframe, we haven't set the src in the HTML. We want to prevent situations where the iframe could load before we've attached a load event handler to it.

We start by adding a message event listener to the container window:

```
<script>
  window.addEventListener('message', (event) => {
    if (event?.data) {
      const { type, data } = JSON.parse(event?.data);
      switch (type) {
        case 'init': {
          console.log('Parent received init message');
          return;
        }
      }
    }
  });
</script>
```

Next, we'll select the payment capture iframe and add a load event listener to the iframe element. Our handler will send an init message with some data to the iframe element's contentWindow:

```
<script>
  // no change to the message listener
  const $paymentCaptureIframe = document.querySelector
    ('#payment-capture');
```

```
  // on iframe load, we'll send a message
  $paymentCaptureIframe.addEventListener('load', () => {
    $paymentCaptureIframe.contentWindow.postMessage(
      JSON.stringify({ type: 'init', data: { price: 20000,
        currency: 'EUR' } }),
    );
  });
</script>
```

Finally, we can set the `iframe` element's `src` attribute so that it loads:

```
<script>
  // no change to message and iframe load listeners
  $paymentCaptureIframe.setAttribute(
    'src',
    new URL('/frame-content.html', window.location.origin),
  );
</script>
```

We now need to implement the `frame-content.html` file to receive the message we sent. Our `iframe`, again is mostly a heading and a form with multiple fields. We have `type=hidden` inputs for the price and currency, as well as text inputs for the card number, expiry date, and **card verification value (CVV)** code. We also include a `Messages` section to illustrate which messages are being sent and received by the iframe:

```
<h2>Payment iframe</h2>
<form>
  <input type="hidden" name="price" />
  <input type="hidden" name="currency" />
  <div>
    <label for="cardnumber">Card Number</label>
    <input required="required" name="cardnumber"
      id="cardnumber" type="text" />
  </div>
  <div>
    <label for="cardexpiry">Expiry Date</label>
    <input required="required" name="cardexpiry"
      id="cardexpiry" type="text" />
  </div>
  <div>
    <label for="cardcvv">CVV</label>
    <input name="cardcvv" id="cardcvv" type="text" />
  </div>
</form>
```

```
<div>
  <h3>Messages</h3>
  <pre><code></code></pre>
</div>
```

In order to handle messages from the parent frame, we'll add a `message` event listener. It stores all received messages in the `pre` code element we defined earlier.

If the `event.data.type` is `init`, we set the value of our `price` and `currency` inputs:

```
<script>
  window.addEventListener('message', async (event) => {
    document.querySelector('pre code').innerText +=
      `Received: ${event.data}\n`;
    const { type, data } = JSON.parse(event.data);
    switch (type) {
      case 'init': {
        document.querySelector('[name=price]').value =
          data.price;
        document.querySelector('[name=currency]').value =
          data.currency;
        return;
      }
    }
  });
</script>
```

Finally, we send an `init` message when our script finished running. We use `window.parent.postMessage` to achieve this:

```
<script>
  // no change to the message event listener
  const initMessage = JSON.stringify({ type: 'init' });
  document.querySelector('pre code').innerText += `Sent:
    ${initMessage}\n`;

  window.parent.postMessage(initMessage);
</script>
```

With this code in place, when we load the `frame-parent.html` file in a browser, we see the following. The `iframe` has sent an `init` message and received one as well.

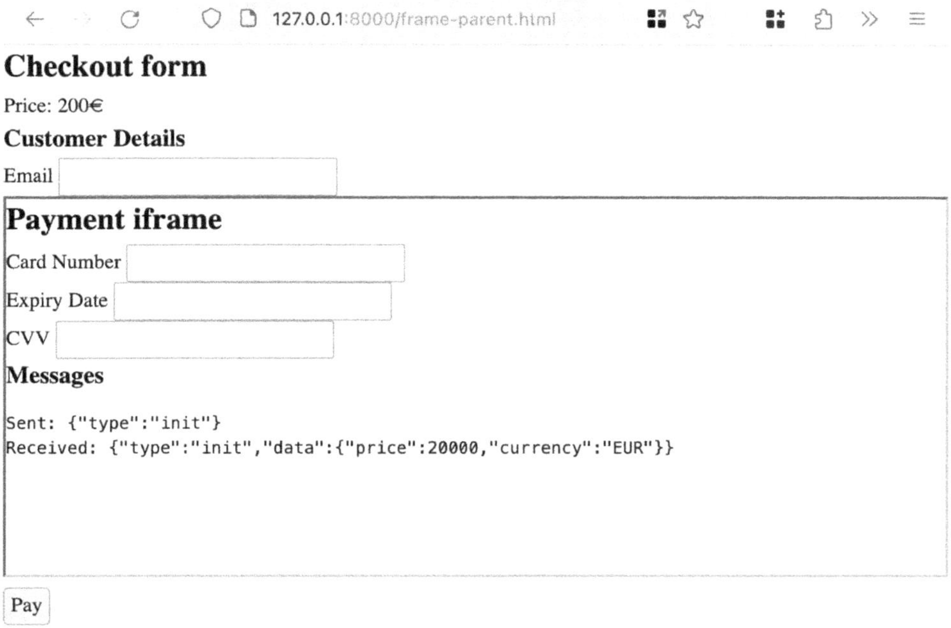

Figure 8.7: Container and iframe contents in their initial state

When we submit the container, we'll want to ensure the card details are retrieved by the `iframe` and passed back to the container. These details will be encrypted by the `iframe` (which, in our scenario, will be served from a domain from the payment service provider) before being sent to the parent document.

The following diagram details the expected interactions.

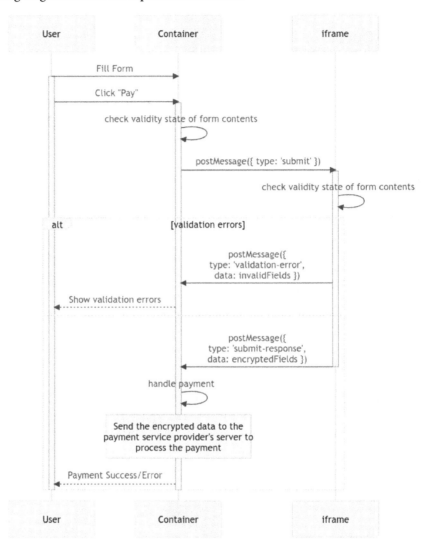

Figure 8.8: Container and iframe communication sequence diagram during user interaction

The key change we have to make to the container is to listen for a submit event on the form element. We then send a message with type="submit" to the iframe:

```
<script>
  // no message to iframe initialization code
  document.querySelector('form').addEventListener('submit',
```

```
    (event) => {
      event.preventDefault();
      $paymentCaptureIframe.contentWindow.postMessage(
        JSON.stringify({
          type: 'submit',
        }),
      );
    });
</script>
```

The iframe receives the message and we'll need to extend our message event handler to react to the submit message:

```
<script>
  window.addEventListener('message', async (event) => {
    // no change outside of the switch
    switch (type) {
      case 'submit': {
        $form = document.querySelector('form');

        const isFormValid = $form.reportValidity();
        if (!isFormValid) {
          const fields = [...$form];
          const invalidFields = fields
            .filter((f) => f.type !== 'hidden' &&
              !f.validity?.valid)
            .map((f) => f.name);
          const message = JSON.stringify({
            type: 'validation-error',
            data: invalidFields,
          });
          window.parent.postMessage(message);
          document.querySelector('pre code').innerText +=
            `Sent: ${message}\n`;
          return;
        }
      }
      // no change to other "case" statements
    }
  });
</script>
```

Now that we've implemented a new `iframe` to container "validation error" message, we need to handle that message type in `frame-parent.html`. In this case, we've already done everything that's necessary in the `submit` form event handler (which calls `preventDefault()`), so we'll simply log out the message contents:

```
<script>
  window.addEventListener('message', (event) => {
    if (event?.data) {
      const { type, data } = JSON.parse(event?.data);
      switch (type) {
        // no change to other "case" statements
        case 'validation-error': {
          console.log('Received message', type, data);
          return;
        }
      }
    }
  });
```

We can now attempt to click **Pay** with the card number and card expiry empty, which yields the following state, where both fields are in the `validation-error` message as received by the frame parent.

Figure 8.9: If you click Pay with the card number and expiry date missing, a
validation-error message between the iframe and the container occurs. Also,
note the HTML validation error message for the card number field

If we then enter the card number but still don't enter the expiry date, then the second `validation-error` message only contains the `cardexpiry` field:

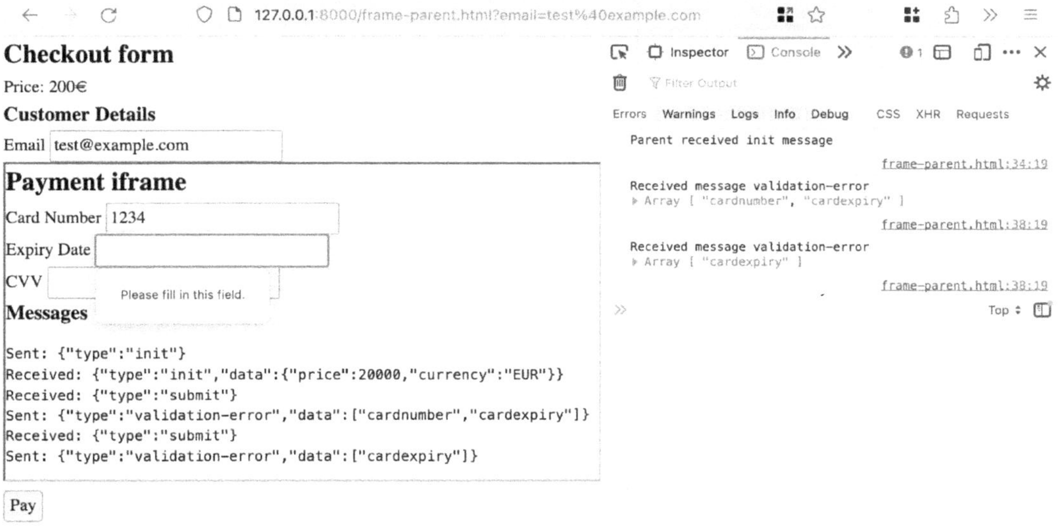

Figure 8.10: If you click Pay with the card number entered and the expiry date missing, a validation-error message between the iframe and the container occurs. Also, note the HTML validation error message for the card expiry field

We need a function to take a string (in our case containing a JSON-encoded JavaScript object) and turn it into a base64-encoded ciphertext (encrypted string in base64 format). The payment service provider usually would manage this encryption, so we wouldn't need this function or to fetch `public-key.json` to enable RSA-OAEP (asymmetric) encryption in the browser.

The code will convert the string to a Uint8Array, fetch a public key, and import it in order to use it with `crypto.subtle.encrypt`. We encrypt the message string (that was converted to a Uint8Array). This yields an `ArrayBuffer` that we encode to base64 by creating a `Uint8Array` object with our data, converting it back to an array and for each character, looking up the relevant character code. Once we have a string containing the character codes, we base64-encode it:

```
<script>
  async function encryptToBase64(message) {
    const msgUint8 = new TextEncoder().encode(message);
    const publicKeyExport = await fetch
      ('./public-key.json').then((res) =>
      res.json(),
    );
    const publicKey = await crypto.subtle.importKey(
      'jwk',
```

```
      publicKeyExport,
      {
        name: 'RSA-OAEP',
        hash: 'SHA-256',
      },
      true,
      ['encrypt'],
    );
    const encryptedBuffer = await crypto.subtle.encrypt(
      {
        name: 'RSA-OAEP',
      },
      publicKey,
      msgUint8,
    );
    return btoa(
      return [...new Uint8Array(encryptedBuffer)]
        .map((el) => String.fromCharCode(el))
        .join(''),
    );
  }
</script>
```

We can now use the `encryptToBase64` function in our `type=submit` message-handling code. Once the validation passes, we'll serialize the data using `FormData`, `FormData().entries()`, and `Object.fromEntries`. We stringify it before encrypting it to a base64 ciphertext.

Finally, we send a `type=submit-reponse` message to the container document with the encrypted string as the payload:

```
<script>
  window.addEventListener('message', async (event) => {
    // no change outside of the switch
    switch (type) {
      case 'submit': {
        $form = document.querySelector('form');
        // no change to form validity validation
        const data = new FormData($form);
        const serializableData = Object.fromEntries
          (data.entries());
        const message = JSON.stringify({
          type: 'submit-response',
          data: await encryptToBase64
            (JSON.stringify(serializableData)),
```

```
      });
      window.parent.postMessage(message);
      document.querySelector('pre code').innerText +=
        `Sent: ${message}\n`;
      return;
    }
    // no change to other "case" statements
  }
  });
</script>
```

We now need to handle the `type=submit-response` message in `iframe-parent.html`. Again, we're just extending our `switch(type)` statement with an additional case for `submit-response`. We'll log some messages, including the `event.data` and extract the values from the container `form` element using `FormData().entries()` and `Object.fromEntries()`. At this point, we could send the `event.data` and the container form data to a backend endpoint to complete the transaction:

```
<script>
  window.addEventListener('message', (event) => {
    if (event?.data) {
      const { type, data } = JSON.parse(event?.data);
      switch (type) {
        // no change to other "case" statements
        case 'submit-response': {
          console.log('received submit-response');
          console.log(event.data);
          const formData = new FormData
            (document.querySelector('form'));
          Const pageData = Object.fromEntries
            (formData.entries());
          return;
        }
      }
    }
  });
```

We can see this in action when we fill out the customer email and the payment details form and click **Pay**:

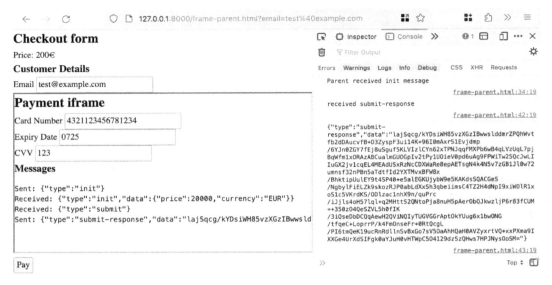

Figure 8.11: When the form is complete and Pay has been clicked, the iframe
receives the submit message and sends back a submit-response message with
an encrypted base64 string, which the container page receives

In order to avoid untrusted frames from sending arbitrary messages, we should check `event.origin` against an allowlist.

We'll add this as a guard clause in the `window.addEventListener` in both `frame-parent.html` and `frame-content.html`. We'll check the message origins against our allowlist. If the `event.origin` is not in the allowlist, we log a warning and discard the message by doing an early return.

In the case of local development, the origin will be `'http://127.0.0.1:8000'` for both interactions. As throughout this section, in a production use case, the allowed origin for receiving messages in the container window (`frame-parent.html`) would be a domain managed by the payment service provider. The `frame-content.html` equivalent would be what the payment service provider hosts, so the allowed domain would be the URL of the container application:

```
<script>
  // handle receiving messages from iframe -> parent
  const allowedMessageOrigins = ['http://127.0.0.1:8000'];
  window.addEventListener('message', (event) => {
    if (!allowedMessageOrigins.includes(event.origin)) {
      console.warn(
        `Dropping message due to non-allowlisted origin
```

```
        ${event.origin}`,
      event,
    );
    return;
  }
  // no change to the rest of the message handler
});
</script>
```

We've now seen how to implement secure messaging between an iframe and the page that contains it. Next, we'll recap on event listener performance anti-patterns.

Event listener performance antipatterns

Event listener performance antipatterns change over time. For example, when Internet Explorer support was broadly required due to its market share, adding event listeners to DOM nodes and subsequently deleting the nodes would not clean up the event listeners, causing memory leaks. This doesn't occur anymore in modern browsers.

An event listener antipattern that is often caught by the Lighthouse page performance auditing tool is scroll event listeners that aren't set to be passive. Passive event listeners are more performant because event.preventDefault() doesn't intercept and stop the event's default behavior. This allows browsers to set the event listener to be non-blocking since the listener can't act on the event.

Making an event listener passive simply involves passing { passive: true } as the third parameter to addEventListener():

```
document.addEventListener(
  'scroll',
  (event) => {},
  { passive: true }
);
```

Another antipattern is to forgo using debounce or throttle on the event listener handler for high-volume events (scroll is a good example). We covered how to implement debounce and throttle in *Chapter 7, Asynchronous Programming Performance Patterns*, in the *Throttling, debouncing and batching asynchronous operations* section.

The final antipattern is solved by event delegation. At some amount of DOM nodes and event listeners, adding one event listener per potential target starts causing performance implications. Luckily, event delegation solves this problem. It allows us to attach one event listener per event type while maintaining the ability to handle each target differently.

We've now covered event listener performance antipatterns to keep an eye out for and how to remediate them.

Summary

In this chapter, we've covered advanced event-driven programming patterns to keep a JavaScript code base performant and secure when handling large numbers of events and event listeners.

Event delegation is useful to ensure that the number of event listeners doesn't grow with the number of DOM nodes in a client-side application where elements are inserted and removed dynamically.

Patterns for secure frame messaging mean we're able to orchestrate `iframe` initialization and bidirectional communication between an `iframe` and its parent document.

Finally, we covered common event listener performance antipatterns to avoid the common pitfalls of event listener-heavy code bases.

Now that we're familiar with advanced event-driven programming patterns in JavaScript, in the next chapter, we'll cover lazy-loading and code-splitting to maximize the performance of JavaScript applications.

9

Maximizing Performance – Lazy Loading and Code Splitting

In order to maximize the performance of a JavaScript application, reducing the amount of unused JavaScript being loaded and interpreted is key. The techniques that can be brought to bear on this problem are called **lazy loading** and **code splitting**. Lazy loading and code splitting allows parts of the JavaScript to be loaded on demand as required. This is in contrast to being downloaded on page load and can greatly reduce the amount of unused JavaScript being loaded and interpreted.

We'll cover the following topics in this chapter:

- The dynamic import syntax and how Vite can automatically code-split based on the syntax

- Route-based code splitting with Next.js and how to read the Bundle Analyzer reports

- How to use `next/dynamic` and `react-intersection-observer` to load JavaScript and React components on different user interactions

By the end of this chapter, you'll be able to identify and leverage lazy loading and code splitting in a variety of scenarios and applications.

Technical requirements

You can find the code files for this chapter on GitHub at `https://github.com/PacktPublishing/Javascript-Design-Patterns`

Dynamic imports and code splitting with Vite

Dynamic imports in JavaScript refer to the usage of the `import()` syntax to import a module. Unlike the `import Something from './my-module.js'` declarative syntax, `import()` is more akin to a function that returns a promise. For example, we could rewrite the original import as `const Something = await import('./my-module.js')`.

The "dynamic" part of the import is that it doesn't have to be done at module evaluation time; it's done as part of the execution of the code. This is useful when paired with code splitting – which we'll define now – since it means that we can avoid loading and evaluating some JavaScript code until it's needed.

Code splitting is a technique whereby code is built into multiple files (also known as "chunks" or "bundles") instead of a single file. Code splitting is useful to avoid loading all the code up front. Instead, when paired with dynamic imports, code is split into multiple files such that different parts of it are loaded only when necessary. This means that there's a lower up-front cost to the JavaScript load, parse, and compile cycle.

The Vite build tool supports code splitting at dynamic import boundaries.

Given a simple document as follows, which has an `id="app"` div and references a `main.js` file, Vite can run a build as long as `main.js` exists:

```
<div id="app"></div>
<script src="./main.js" type="module"></script>
```

We'll have two modules now: `main.js`, which is the entry point that Vite will reference, and our code will import the `dynamic.js` module.

`main.js` will inject `'Hello from main.js'` into our app div. It will then proceed to dynamically load the `dynamic.js` module and set the contents of the app div to the output of the `hello` function as exported by `dynamic.js`:

```
document.querySelector('#app').textContent = 'Hello from main.
js';
const { hello } = await import('./dynamic.js');
document.querySelector('#app').textContent = hello();
```

Here is a simpler `dynamic.js` implementation of the `hello` function:

```
export function hello() {
  return 'Hello from dynamic.js';
}
```

When running the Vite dev server using `npx vite`, we can see that the dynamically imported `hello` function contents are displayed on the page. Notice that `dynamic.js` is loaded as a separate request to `main.js`; that is code splitting at play.

Figure 9.1: "Hello from dynamic.js" on the page with network requests,
including a request specifically for dynamic.js

This pattern can be useful to defer loading JavaScript until it's required – for example, if we want to add client-side tracking of button clicks using `fetch` requests.

We have the following HTML, with two buttons that have a `data-track` property:

```
<!-- no change to app div -->
<div>
  <button data-track="button-click">With tracked click
  </button>
  <button data-track="alt-button-click">Other tracked click
  </button>
</div>
<!-- to change to script -->
```

We'll add a `trackInteraction.js` module with a `trackInteraction` function, which will use `fetch` and the `POST` HTTP method to send interaction data to `jsonplaceholder`. If this were a live implementation, we could realistically replace `jsonplaceholder` with Google Analytics or another equivalent service that exposes a client-side JavaScript accessible endpoint:

```
export function trackInteraction(page, type = 'click') {
  return fetch
    ('https://jsonplaceholder.typicode.com/posts', {
    method: 'POST',
    body: JSON.stringify({
      type,
      page,
    }),
    headers: {
      'Content-type': 'application/json; charset=UTF-8',
    },
```

```
    }).then((response) => response.json());
}
```

Now, the `trackInteraction` module has nothing to do with the page functionality so we want to avoid loading it until it's needed.

In this case, we'll attach a click event listener to each element that has a `data-track` attribute. Only when the listener is triggered does the `import('./trackInteraction.js')` statement run:

```
// no change to rest of main.js

document.querySelectorAll('[data-track]').forEach((el) => {
  el.addEventListener('click', async (event) => {
    const page = window.location.pathname;
    const type = event.target.dataset?.track;
    const { trackInteraction } = await import
      ('./trackInteraction.js');
    const interactionResponse = await trackInteraction
      (page, type);
    console.assert(
      interactionResponse.type === type &&
        interactionResponse.page === page,
      'interaction response does not match sent data',
    );
  });
});
```

If we load the Vite dev server and click the **With tracked click** button and the **Other tracked click** button once and then the **With tracked click** button once again, we'll get the following network requests:

Figure 9.2: Network requests after clicking the "With tracked click," "Other tracked click," and "With tracked click" buttons in sequence

On the first click of either button, the `trackInteraction.js` file is loaded and then a `fetch` request is triggered. On subsequent clicks, `trackInteraction.js` is already loaded so the `fetch` requests to `jsonplaceholder` are the only network requests we see.

Note that each `POST` request to `jsonplaceholder` is preceded by an `OPTIONS` request due to browser **cross-origin resource sharing** (**CORS**). The `OPTIONS` response includes `Access-Control-Allow-`... headers that allow our origin and method.

We've now seen what dynamic imports in JavaScript look like and how Vite automatically code splits dynamic imports, which allows us to only load modules that are required "just in time," thereby allowing us to reduce the upfront JavaScript load/parse/evaluation cost.

Next, we'll cover route-based code splitting in Next.js and how to inspect generated chunks with the Next.js Bundle Analyzer plugin.

Route-based code splitting and bundling

Let's begin by defining a **route** in a general web application context and then in a Next.js context.

In a web application, a route comes from the **router** concept; in simple terms, it's an entry in the router. An entry in the router mechanism can take multiple shapes – for example, in an nginx/Apache/Caddy web server setup, we can have a path to file forwarding or a wildcard forwarding approach. In backend MVC frameworks such as Ruby on Rails, Laravel (PHP), and Django (Python), a route associates a request path to the specific code to be run. The *request path to code to be run* concept also applies to Node.js backend applications using libraries such as Express, Koa, Fastify, and Adonis.js.

Let's now see how the *route* concept is used in the Next.js filesystem router.

A minimal Next.js project as initialized with `create-next-app` is laid out as follows. Each file in the `pages` directory corresponds to a route. For example, `index.js` is used to render the `/` path of the application. If we had an `about.js` or `about/index.js` file, that would be used to render the `/about` path of the application:

```
.
├── components
├── next.config.js
├── package.json
├── pages
│   └── index.js
└── public
```

We defined code splitting in the previous section, *Dynamic imports and code splitting with Vite*. Since a core Next.js feature is the router, it can do what's called **route-based code splitting**, which is automatic code splitting based on a route or page contents.

A naive route-based code-splitting approach would be to create completely separate sets of bundles for each route. In the context of a React or Next.js application, this is inefficient since we would end up with shared libraries (for example, React and Next.js) in each of the per-page bundles.

What Next.js can do in this case is identify shared code and classify it as `First Load JS shared by all`.

This is the sample build output:

```
+ First Load JS shared by all                79.9 kB
  ├ chunks/framework-cc1b0d6c55d15cb9.js     45.3 kB
  ├ chunks/main-7c6ad51e94ec3ff5.js          32.8 kB
  ├ chunks/pages/_app-db3a4be757903450.js    205 B
  └ chunks/webpack-8850afd7843acaaa.js       1.55 kB
```

We can add the Next.js Bundle Analyzer to check the contents of each chunk:

```
npm install --save @next/bundle-analyzer
```

Then, we can configure `next.config.js` to use it. In our case, it looks as follows:

```
const nextConfig = {
  // no changes to this config
};
const withBundleAnalyzer = require('@next/bundle-analyzer')({
  enabled: process.env.ANALYZE === 'true',
});

module.exports = withBundleAnalyzer(nextConfig);
```

To use the bundle analyzer, we can add an `analyze` script to our `package.json file`:

```
{
  "//": "// no change to other properties",
  "scripts": {
    "//": "// no change to other scripts",
    "analyze": "cross-env ANALYZE=true next build"
  }
}
```

This can be run with npm run analyze. Its shell output is the same as npm run build but it opens a browser window with the bundle analysis file – for example, .next/analyze/client. html:

```
npm run analyze

> next-route-based-splitting@0.1.0 analyze
> cross-env ANALYZE=true next build

 ✓ Linting and checking validity of types
Webpack Bundle Analyzer saved report to /next-route-based-splitting/.
next/analyze/nodejs.html

No bundles were parsed. Analyzer will show only original module sizes
from stats file.

Webpack Bundle Analyzer saved report to /next-route-based-splitting/.
next/analyze/edge.html
Webpack Bundle Analyzer saved report to /next-route-based-splitting/.
next/analyze/client.html

 ✓ Creating an optimized production build
 ✓ Compiled successfully
 ✓ Collecting page data
 ✓ Generating static pages (3/3)
 ✓ Collecting build traces
 ✓ Finalizing page optimization
```

We can use this to inspect the contents of the `shared` JavaScript in the `framework`, `main`, `pages/_app`, and `webpack` chunks as well as page-specific chunks:

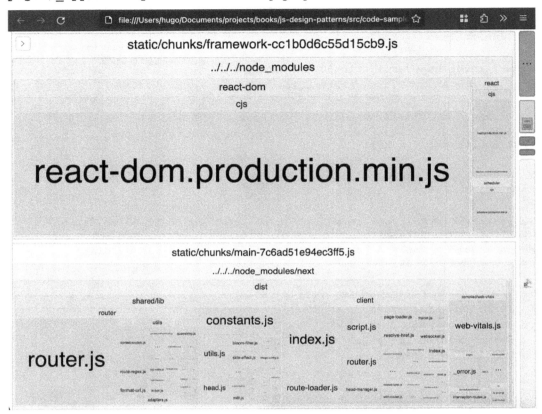

Figure 9.3: The client.html Bundle Analyzer output in the browser

The `framework` bundle includes the following packages from `node_modules`: `react`, `react-dom`, and `scheduler`. Meanwhile, the `main` bundle includes `next` and its submodules such as `shared/lib`, which includes a large `router` chunk, or `next/client`, which is the client-side section of Next.js. Also, it is harder to see in the preceding screenshot, but `main` includes `@swc/helpers/esm`, which is probably an artifact of the Next.js build using the SWC compiler.

We've now seen how Next.js supports route-based code splitting and how to inspect the contents of the Next.js-generated bundles using the Next.js Bundle Analyzer report. Next, we'll see dynamic import patterns to load additional JavaScript under different element visibility and interaction conditions.

Loading JavaScript on element visibility and interaction

In this section, we'll look at four different scenarios where dynamic or lazy loading of React components and JavaScript modules can be applied in the context of a Next.js application.

The first instance will be whether the component is in the component tree or not – in other words, whether it's considered to be rendered or not. Next, we'll look at dynamic imports based on user interaction. We'll also cover how to handle an interaction that potentially requires a dynamic import of a JavaScript resource. Finally, we'll show how to dynamically load a React component when an element is visible in the viewport.

Next.js provides a `dynamic` utility (see the documentation at `https://nextjs.org/docs/pages/building-your-application/optimizing/lazy-loading`) that allows us to lazily and dynamically load a React component.

In our case, we have a `components/Hello.jsx` component with a `Hello` component that is a named export:

```
import React from 'react';

export function Hello() {
  return <>Hello</>;
}
```

We can dynamically load it using `dynamic()` and `import()`. Due to `Hello` being a named export, we need to extract the `Hello` property of the `import()` promise using `.then()`. We set `ssr: false` to showcase how `next/dynamic` allows us to control whether a dynamically loaded component is included in the server-rendered output or not:

```
import React from 'react';
import dynamic from 'next/dynamic';
const DynamicClientSideHello = dynamic(
  () => import('../components/Hello.jsx').then(({ Hello })
    => Hello),
  { ssr: false },
);
export default function Index() {
  return (
    <>
      <h1>Next.js route-based splitting and component lazy
        loading</h1>
      <DynamicClientSideHello />
    </>
  );
}
```

By using npm run analyze as configured in the *Route-based code splitting and bundling* section (using the @next/bundle-analyzer module), we can inspect the contents of the chunks/pages/index chunk; you'll note that Hello.jsx is in a different chunk.

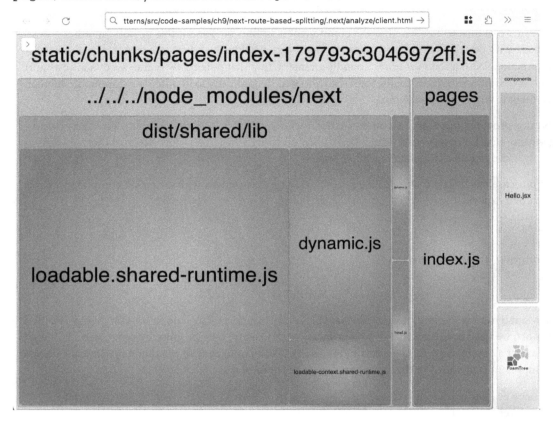

Figure 9.4: Bundle analyzer filtered to "chunks/pages/index" and the chunk containing Hello.jsx

When we run the Next.js dev server using next dev and load up the / path, we see the following page and network requests. _next/static/chunks/components_Hello_jsx.js is loaded last and separately to _next/static/chunks/pages/index.js, which means that we are in fact doing a dynamic load of the Hello.jsx component.

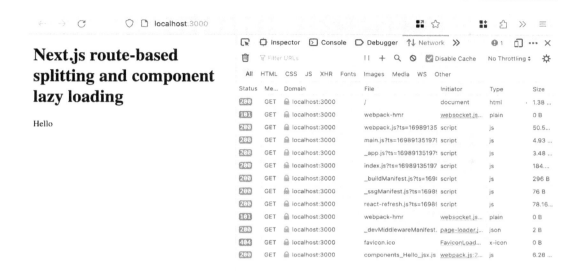

Figure 9.5: Dynamic loading of the Hello.jsx page contents and Network tab

We'll now showcase using next/dynamic inside of the Index component based on the component state.

Our example is a *Terms and Conditions* selector that allows the user to select between three options: **None**, **Short**, and **Long**. **None** will be handled by a NoRender component (which simply returns null), and **Short** and **Long** will dynamically load a component to display.

We'll start by adding a components/TermsAndConditionsShort.jsx component, which contains an h3 element and a single paragraph of content:

```
import React from 'react';

export function TermsAndConditions() {
  return (
    <>
      <h3>Terms and Conditions Short</h3>

      <p>{/* Terms and Conditions Content */}</p>
    </>
  );
}
```

We'll also add a `components/TermsAndConditionsLong.jsx` component, which contains the same h3 and content but has five paragraphs of content instead of one:

```jsx
import React from 'react';
export function TermsAndConditions() {
  return (
    <>
      <h3>Terms and Conditions Long</h3>
      <p>{/* Terms and Conditions Content */}</p>
      <p>{/* Terms and Conditions Content */}</p>
      <p>{/* Terms and Conditions Content */}</p>
      <p>{/* Terms and Conditions Content */}</p>
      <p>{/* Terms and Conditions Content */}</p>
    </>
  );
}
```

Finally, we'll add a `select` field with relevant `option` values (None, Short, and Long) to `pages/index.js`. We'll use `useState` to keep track of the currently selected option:

```jsx
import React, { useState } from 'react';
export default function Index() {
  const [selectedTermsAndConditions,
    setSelectedTermsAndConditions] =
    useState('None');

  return (
    <>
      {/* no changes to rest of the returned JSX */}
      <div>
        <label htmlFor="termsAndConditionsType">
          Terms and Conditions selector:
        </label>
        <select
          id="termsAndConditionsType"
          onChange={(e) => setSelectedTermsAndConditions
            (e.target.value)}
        >
          <option value="None">None</option>
          <option value="Short">Short</option>
          <option value="Long">Long</option>
        </select>
      </div>
```

```
      </>
   );
}
```

Finally, we'll add a `NoRender` component and, based on `selectedTermsAndConditions`, either render `NoRender` or the dynamically loaded `TermsAndConditions` component:

```
import React, { useState } from 'react';
const NoRender = () => null;
export default function Index() {
  // no change to useState
  const TermsAndConditions = ['Short', 'Long'].includes(
    selectedTermsAndConditions,
  )
    ? dynamic(() =>
        import(
          `../components/TermsAndConditions$
            {selectedTermsAndConditions}.jsx`
        ).then(({ TermsAndConditions }) =>
          TermsAndConditions),
      )
    : NoRender;
  return (
    <>
      {/* no changes to rest of the returned JSX */}
      <div>
        {/* no change to label or select */}
        <hr />
        <TermsAndConditions />
      </div>
    </>
  );
}
```

When we run the next dev server and load the index page, we initially see the **None** state selected and no dynamic imports, apart from the existing `Hello.jsx` one from the previous example.

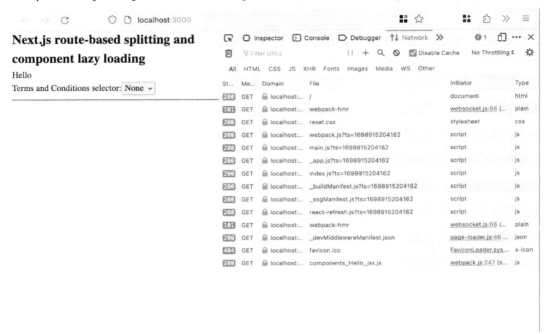

Figure 9.6: Terms and conditions selector initial state with None selected;
therefore, no dynamic imports apart from the existing Hello.jsx one

On selection of **Short**, we note that the relevant heading and paragraph are displayed and we have an additional request that loaded `_next/static/chunks/components_ TermsAndConditionsShort_jsx.js`.

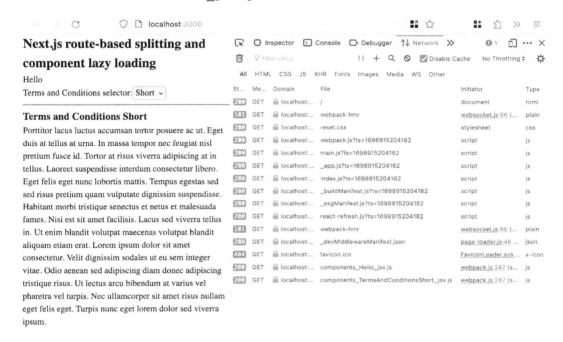

Figure 9.7: Terms and conditions selector when Short is selected;
TermsAndConditionsShort.jsx has been dynamically loaded and is displayed

When we select **Long**, we note that the relevant heading and paragraph are displayed and we have an additional request that loaded `/_next/static/chunks/components_TermsAndConditionsLong_jsx.js`.

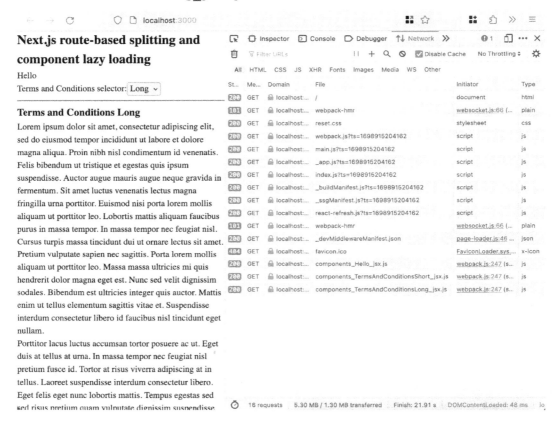

Figure 9.8: Terms and conditions selector when Long is selected;
TermsAndConditionsLong.jsx has been dynamically loaded and is displayed

We can also look at the Bundle Analyzer's `client.html` output using `npm run analyze`; the following has been filtered to the relevant chunks to illustrate how `TermsAndConditionsShort` and `TermsAndConditionsLong` are not included in `chunks/pages/index.js`. There are three "dynamic" chunks (which correlates with our findings from the network requests we observe in the browser): one for `components/Hello.jsx`, one for `components/TermsAndConditionsShort.jsx`, and one for `components/TermsAndConditionsLong.jsx`.

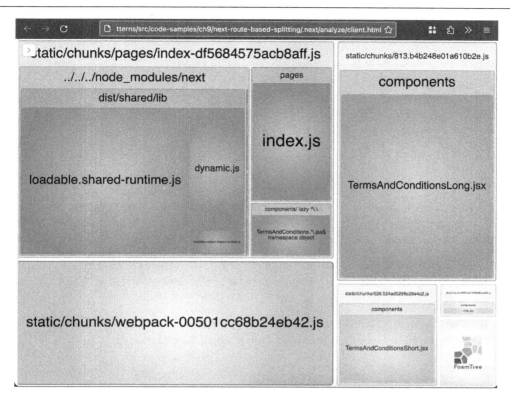

Figure 9.9: Bundle Analyzer output for the page chunk as well as the dynamic chunks (which include the TermsAndConditionsShort and TermsAndConditionsLong components)

We've now seen how `dynamic` can be used in response to a user action to dynamically load content based on user-provided data. Next, we'll revisit dynamic imports of a JavaScript resource (as opposed to React components) while handling a user action in the context of a Next.js application.

We'll start with a new component, `TermsAndConditionsLongScroll.jsx`, which is functionally the same as `TermsAndCondtionsShort.jsx` or `TermsAndCondtionsLong.jsx` but with 10 paragraphs:

```
import React from 'react';

export function TermsAndConditions() {
  return (
    <>
      <h3>Terms and Conditions Long Scroll</h3>
      <p>{/* Terms and Conditions Content */}</p>
      <p>{/* Terms and Conditions Content */}</p>
      <p>{/* Terms and Conditions Content */}</p>
```

```
      <p>{/* Terms and Conditions Content */}</p>
      <p>{/* Terms and Conditions Content */}</p>
      <p>{/* Terms and Conditions Content */}</p>
      <p>{/* Terms and Conditions Content */}</p>
      <p>{/* Terms and Conditions Content */}</p>
      <p>{/* Terms and Conditions Content */}</p>
      <p>{/* Terms and Conditions Content */}</p>
    </>
  );
}
```

We'll now add a form at the bottom of the page to accept the terms and conditions. We have a long form so it's nice to be able to go directly to the bottom. To this end, we add a button that, on click, scrolls us to the input checkbox element using a ref.

In our *scroll-to-bottom* handler, we ensure that smooth scrolling is available (some older Safari versions don't natively support it) by conditionally importing the `scroll-behavior-polyfill` package if `scrollBehavior` is not detected.

Finally, we scroll using the `scrollTargetRef.current.scrollIntoView()` function. `scrollTargetRef` is attached to the checkbox input using the `ref` property:

```
import React, { useRef } from 'react';

export function TermsAndConditions() {
  const scrollTargetRef = useRef();

  async function handleScroll() {
    if (!('scrollBehavior' in document.
      documentElement.style)) {
      await import('scroll-behavior-polyfill');
    }
    if (scrollTargetRef.current) {
      scrollTargetRef.current.scrollIntoView({
        behavior: 'smooth',
        block: 'end',
      });
    }
  }

  return (
    <>
      {/* no change to heading */}
      <button onClick={handleScroll}>Scroll to button
```

```
        </button>
      {/* no changes to paragraphs */}
      <hr />
      <label htmlFor="accept">
        <input
          id="accept"
          name="acceptTerms"
          type="checkbox"
          ref={scrollTargetRef}
        />
        Accept Terms and Conditions
      </label>
    </>
  );
}
```

Back in pages/index.js, we'll allow **LongScroll** to be selected (as a new option) and to be dynamically imported:

```
// no changes to imports and definitions outside of Index
export default function Index() {
  // no changes to useState to maintain select state

  const TermsAndConditions = ['Short', 'Long',
    'LongScroll'].includes(
    selectedTermsAndConditions,
  )
    ? dynamic(() =>
        import(
          `../components/TermsAndConditions$
            {selectedTermsAndConditions}.jsx`
        ).then(({ TermsAndConditions }) =>
          TermsAndConditions),
      )
    : NoRender;
  return (
    <>
      {/* no change to content outside of select */}

      <div>
        {/* no change to label */}
        <select
          id="termsAndConditionsType"
          onChange={(e) => setSelectedTermsAndConditions
```

```
                (e.target.value)}
          >
            {/* no change to existing options */}
            <option value="LongScroll">LongScroll</option>
          </select>
          <hr />
          <TermsAndConditions />
        </div>
      </>
    );
  }
```

When we run the next dev server, load the index page, and select **LongScroll**, we see the following with a **Scroll to bottom** button and a large amount of paragraph content. Note that there is only one network request for `TermsAndConditionsLongScroll.jsx`.

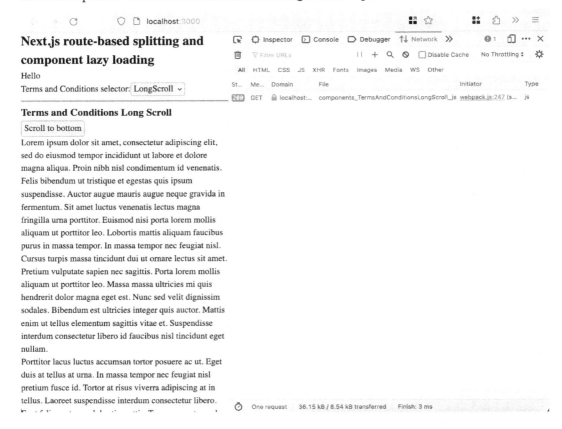

Figure 9.10: TermsAndConditionsLongScroll.jsx selection with dynamic import

In browsers where `behavior:  'smooth'` is supported, when the **Scroll to bottom** button is clicked, no additional JavaScript chunks are loaded and we're scrolled to the checkbox input after the multiple paragraphs.

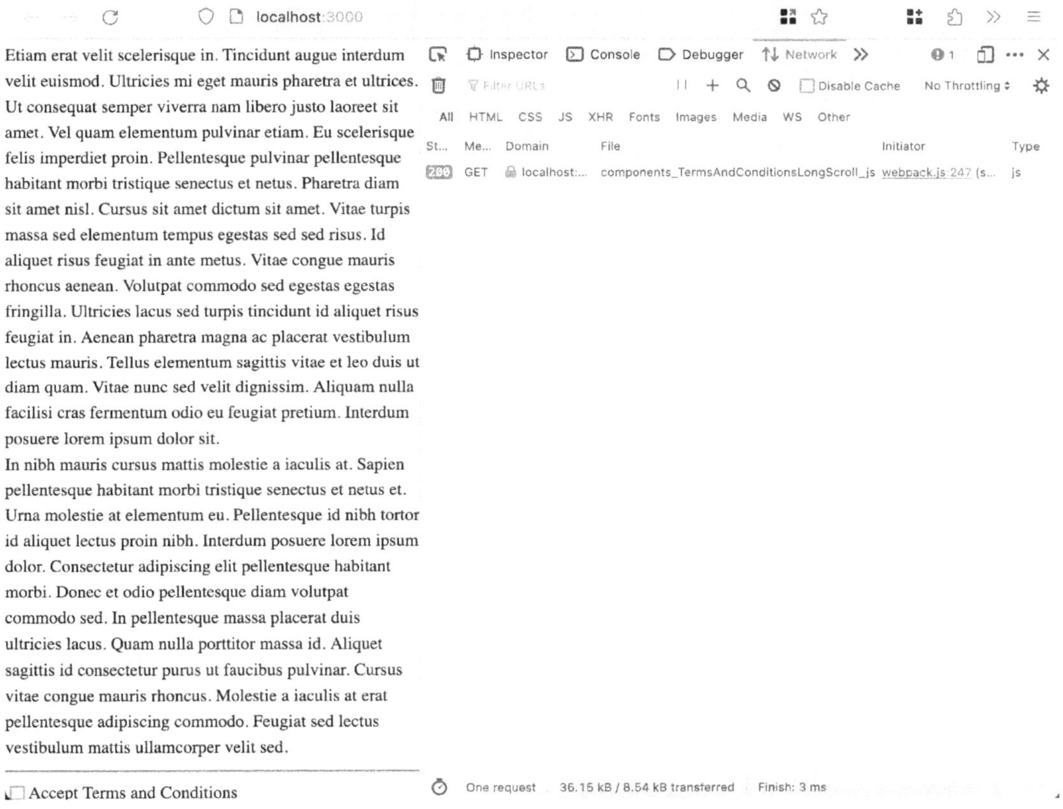

Figure 9.11: TermsAndConditionsLongScroll.jsx selection with dynamic import

On browsers that don't support `behavior: 'smooth'` for scrolling, `scroll-behavior-polyfill` will be loaded allowing for smooth scrolling to the checkbox.

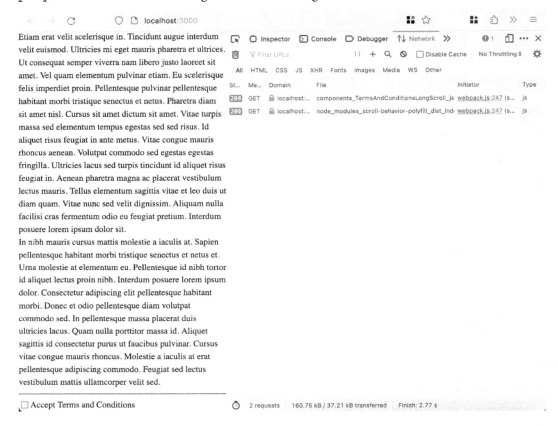

Figure 9.12: TermsAndConditionsLongScroll.jsx selection with dynamic import of the component and of the scroll-behavior-polyfill module

Based on the Bundle Analyzer output (using `npm run analyze` and the `@next/bundle-analyzer` plugin), we can see that there is a chunk that contains `scroll-behavior-polyfill`, along with chunks for `pages/index.js` and one each for `TermsAndConditionsShort.jsx`, `TermsAndConditionsLong.jsx`, and `TermsAndConditionsLongScroll.jsx`.

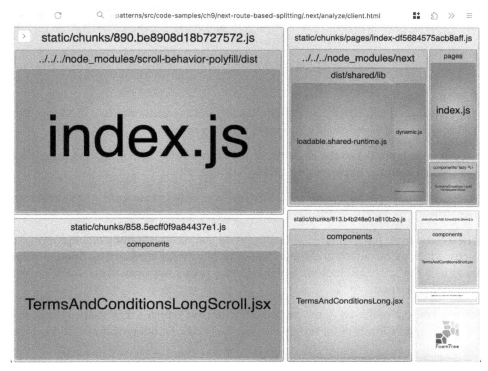

Figure 9.13: Bundle Analyzer output for the pages/index.js chunk as well as
relevant dynamic chunks (TermsAndConditionsShort, TermsAndConditionsLong,
TermsAndConditionsLongScroll, and scroll-behavior-polyfill)

We've now seen that Next.js code splits effectively on native `import()` as well as the provided `dynamic()` utility.

Finally, we'll see how to use `dynamic()` and the `react-intersection-observer` package to dynamically load content when it is visible.

One other variant of a Terms and Conditions form or similar would be to include additional fields that should be captured when the customer accepts the terms.

In this example, we'll add a `components/TermsForm.jsx` component with an input for the user's name and a label for it:

```
import React from 'react';

export default function TermsForm() {
  return (
    <form>
      <label htmlFor="name">Type your name as signature
```

```
      </label>
      <input id="name" type="text" />
    </form>
  );
}
```

Next, we'll want to include it in `components/TermsAndConditionsLongScrollAcceptForm.jsx`. We'll use `dynamic()` to load the `TermsForm` component.

The rest of our code is similar to the end state of the `TermsAndConditionsLongScroll` components, with a heading, 10 paragraphs, and an `accept` input.

The key exception is the import and usage of the `InView` component from `react-intersection-observer`.

The `InView` component has a children render property that receives, among other properties, the `ref` property, which we can attach to elements whose visibility we're interested in. Another property of interest to us is the `inView` Boolean, which tells us whether the element on which we put the `ref` prop is in the viewport.

As the rendered output of the `InView` children function, we return a `div` element to which we attach the `ref` property. Inside of the `div`, we render `TermsForm` but only if `inView` is `true`:

```
import React from 'react';
import dynamic from 'next/dynamic';
import { InView } from 'react-intersection-observer';

const TermsForm = dynamic(() => import('./TermsForm.jsx'));

export function TermsAndConditions() {
  return (
    <>
      <h3>Terms and Conditions Long Scroll Accept Form</h3>
      {/* 10 paragraphs of content */}
      <hr />
      <InView>
        {({ inView, ref }) => <div ref={ref}>{inView &&
          <TermsForm />}</div>}
      </InView>
      <label htmlFor="accept">
        <input id="accept" name="acceptTerms"
          type="checkbox" />
        Accept Terms and Conditions
      </label>
```

```
    </>
  );
}
```

Finally, we need to add `TermsAndConditionsLongScrollAcceptForm` as a selectable option and a dynamically loaded component:

```jsx
// no changes to imports and definitions outside of Index
export default function Index() {
  // no changes to useState to maintain select state

  const TermsAndConditions = [
    'Short',
    'Long',
    'LongScroll',
    'LongScrollAcceptForm',
  ].includes(selectedTermsAndConditions)
    ? dynamic(() =>
        import(
          `../components/TermsAndConditions$
            {selectedTermsAndConditions}.jsx`
        ).then(({ TermsAndConditions }) =>
          TermsAndConditions),
      )
    : NoRender;
  return (
    <>
      {/* no change to content outside of select */}

      <div>
        {/* no change to label */}
        <select
          id="termsAndConditionsType"
          onChange={(e) => setSelectedTermsAndConditions
            (e.target.value)}
        >
          {/* no change to existing options */}
          <option value="LongScrollAcceptForm">
            LongScrollAcceptForm</option>
        </select>
        <hr />
        <TermsAndConditions />
      </div>
```

```
        </ >
    );
}
```

Now, when we run the next dev server and load the index page, `LongScrollAcceptForm` is available. When we select it, the `TermsAndConditionsLongScrollAcceptForm.jsx` component is loaded.

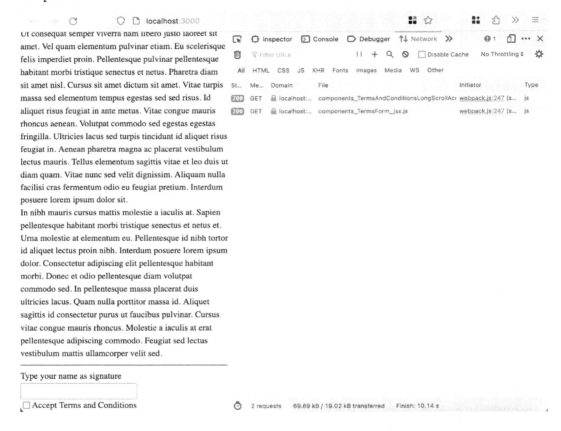

Figure 9.14: LongScrollAcceptForm selected and
TermsAndConditionsLongScrollAcceptForm.jsx dynamically loaded

When `TermsAndConditionsLongScrollAcceptForm` is scrolled to the bottom (to the point where the checkbox is visible), the `TermsForm.jsx` component is dynamically loaded and is shown on the page.

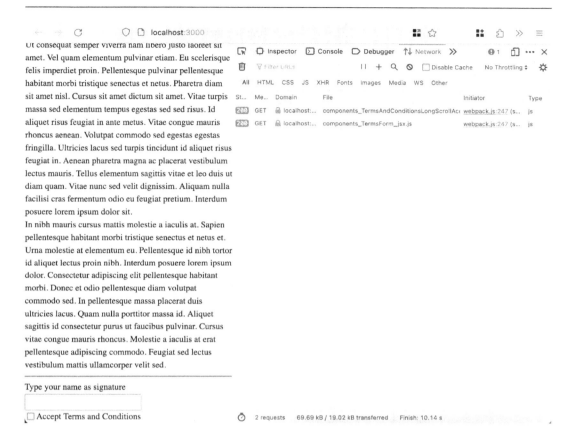

Figure 9.15: TermsAndConditionsLongScrollAcceptForm scrolled to
the bottom and TermsForm.jsx dynamically loaded

We've now seen how to load JavaScript and React components on component visibility and interaction with Next.js.

Summary

In this chapter, we've covered various approaches for maximizing the performance of your JavaScript, React, and Next.js applications with lazy loading approaches and code splitting.

First, we showcased how to use the dynamic import syntax in a Vite-powered setup to cause code splitting and illustrated it by importing additional code only when it's required (during an interaction handler).

Next, we saw how Next.js provides out-of-the-box route-based code splitting while also ensuring modules shared across pages don't get loaded or output more than once. We also delved into how to validate this using the Next.js Bundle Analyzer plugin.

Finally, we covered how to implement different lazy loading scenarios in Next.js: on presence in the component tree, on change caused by user interaction, importing a JavaScript module during an event handler, and lazy loading on an element entering the viewport.

We now know how to leverage lazy loading and code splitting to maximize application load performance. In the next chapter, we'll cover asset-loading strategies and how to execute code off the main thread.

10

Asset Loading Strategies and Executing Code off the Main Thread

There are situations in the life cycle of an application where loading more JavaScript is inevitable. This chapter details techniques to mitigate the impact of such situations. You'll learn about asset loading optimizations such as a script element's `async`, the `defer` attribute, the impact of `type="module"`, and the link element's `rel` (relationship) attribute's `preconnect`, `preload`, and `prefetch` values. Next, you will further optimize script loading using Next.js' `Script` component and its different options. The chapter wraps up with an exploration of reasons to execute JavaScript code off the main thread and an approach to do so.

In this chapter, we'll cover the following topics:

- How to control asset loading more granularly with a script's `async` and `defer` attributes, and links with `preconnect`, `preload`, and `prefetch`
- Further optimization opportunities in Next.js using the `Script` component and its `strategy` prop
- When and how to run code off the main thread via Next.js and Partytown

By the end of this chapter, you'll have the skills to exert more control over asset loading and JavaScript loading and execution in a web context.

Technical requirements

You can find the code files for this chapter on GitHub at `https://github.com/PacktPublishing/Javascript-Design-Patterns`

Asset loading optimization – async, defer, preconnect, preload, and prefetch

When using `script` to load and execute JavaScript, there are HTML attributes of `script` we can use to control the loading and execution.

We can rely on the difference between external scripts and inline scripts; we can also use the `async`, `defer`, and `type="module"` attributes.

We'll start by defining external and inline scripts, then the `async` and `defer` attributes. Finally, we'll look at **classic** and **module** scripts via the `type="module"` attribute.

External scripts use the `src` attribute to point to a separate JavaScript file; for example, what follows is an external script that will load and evaluate `./script.js` when it's encountered:

```
<script src="./script.js"></script>
```

Contrast this with inline scripts, where there is no `src` attribute; instead, the JavaScript code is in the `script` tag contents:

```
<script>
  console.log('inline script');
</script>
```

The default load/execution cycle of scripts is what we call **blocking**. The evaluation of the HTML document will wait for the `script` tag's JavaScript to complete execution.

The `async` and `defer` attributes on the `script` HTML tag can change the behavior of loading and executing scripts.

Adding `async` to a script will mean it's fetched while the rest of the HTML document is parsed. An `async` script will be evaluated as soon as it's loaded. This is a large change to the default document parsing behavior of `script`.

Say we have an `async.js` file that inserts a paragraph with the text `async.js: async script executed`:

```
(() => {
  const node = document.createElement('p');
  node.innerText = 'async.js: async script executed';
  document.body.appendChild(node);
})();
```

Say we also have a `script.js` file that also inserts a paragraph with `script.js: blocking script executed`:

```
(() => {
  const node = document.createElement('p');
  node.innerText = 'script.js: blocking script executed';
  document.body.appendChild(node);
})();
```

Finally, say we have a document that has inline script snippets that also add paragraphs to track their execution before and after two additional `script` tags. One script loads `async.js` with an async attribute on the script, and the second script loads the `script.js` element using the default render-blocking load:

```
<script>
  (() => {
    const node = document.createElement('p');
    node.innerText = 'inline: script 1 executed';
    document.body.appendChild(node);
  })();
</script>
<script src="./async.js" async></script>
<script src="./script.js"></script>
<script>
  (() => {
    const node = document.createElement('p');
    node.innerText = 'inline: script 2 executed';
    document.body.appendChild(node);
  })();
</script>
```

This is displayed as follows in a browser when loaded with an empty cache: the inline script 1 executes first, then `script.js`, then inline script 2, and finally `async.js`. Note how `async.js` was in the document *before* `script.js` but executed after; that's the effect of the `async` attribute:

inline: script 1 executed

script.js: blocking script executed

inline: script 2 executed

async.js: async script executed

Figure 10.1: Inline scripts, external script, and external script with async execution order

Next, we'll see how `defer` affects the loading of a script.

`defer` indicates to the browser that the script should only be loaded *after* the document has been parsed. However, the `DOMContentLoaded` event will not fire until all scripts with the `defer` attribute are loaded and executed.

Say we add a `defer.js` file that will insert a paragraph with `defer.js: defer script executed`, as shown in the following code block:

```
(() => {
  const node = document.createElement('p');
  node.innerText = 'defer.js: defer script executed';
  document.body.appendChild(node);
})();
```

Next, we extend the HTML document from the previous `async` example by adding `<script src="./defer.js" defer></script>` before `<script src="./async.js" async></script>`. This will look as follows:

```
<!-- no change to inline script 1 -->
<script src="./defer.js" defer></script>
<script src="./async.js" async></script>
<script src="./script.js"></script>
<!-- no change to inline script 2 -->
```

When we load this document in a browser, we see the following output where the deferred script adds its paragraph after all the other ones despite being *before* the `async.js`, `script.js`, and inline script 2 in the parse order of the document.

inline: script 1 executed

script.js: blocking script executed

inline: script 2 executed

async.js: async script executed

defer.js: defer script executed

Figure 10.2: Inline scripts, external script, external script with async,
and external script with the defer execution order

Next, we'll see how "module" and "classic" scripts are affected differently by `async` and `defer`.

When a script receives the type of attribute with the `module` value, that script will get interpreted as a JavaScript module. We'll call these "module" scripts, as opposed to "classic" scripts, which don't have a type of attribute.

`type="module"` defers the execution of the script. This means that "module" scripts aren't affected by the `defer` attribute (since that behavior is applied to their execution by default).

The `async` attribute overall has a similar effect on "module" scripts as it does on "classic" scripts, in that the script will be loaded in parallel to document parsing and executed once loading has been completed.

One additional effect of the `async` attribute on "module" scripts is that since JavaScript modules have syntax to denote dependency loading, the module script itself, and once loaded, all the dependencies it loads via the `import` syntax, will be loaded in parallel to the document parsing.

Say we have the following `module.js`, which inserts `module.js: type="module" executed` when it runs:

```
const node = document.createElement('p');
node.innerText = 'module.js: type="module" executed';
document.body.appendChild(node);
```

Say we also have the following `module-async.js`, which inserts `module-async.js: type="module" async executed` when it runs:

```
const node = document.createElement('p');
node.innerText = 'module-async.js: type="module"
  async executed';
document.body.appendChild(node);
```

We add script tags with `type="module"` with an inline module that inserts `inline: type="module" executed`, and module scripts referencing `module.js` and `module-async.js`:

```
<!-- no change to inline scripts -->
<script type="module">
  const node = document.createElement('p');
  node.innerText = 'inline: type="module" executed';
  document.body.appendChild(node);
</script>
<script src="./module-async.js" type="module" async>
  </script>
<script src="./module.js" type="module"></script>
<!-- no change to existing external scripts -->
```

When we load this document in the browser, we see the following. This illustrates that the default load/execution of `type="module"` is deferred since even the inline module script executes after `async` scripts. One point of interest is that `async` on module scripts can make it execute earlier than scripts without `async`. This makes sense since `async` means there's parallel loading and execution is "as soon as available," as opposed to the module script's default execution method, which is `defer`:

← → C ○ ▢ localhost:5173 ⋯ ⬇ ⤴ » ≡

inline: script 1 executed

script.js: blocking script executed

inline: script 2 executed

module-async.js: type="module" async executed

async.js: async script executed

inline: type="module" executed

module.js: type="module" executed

defer.js: defer script executed

Figure 10.3: Inline scripts, external script, external script with async, external script with the defer execution order, inline module script, and external module scripts with async and without

We've now contrasted different characteristics of script load/execution: inline versus external, the impact of `async` and `defer` attributes, and classic versus module. The following diagram recapitulates the execution order:

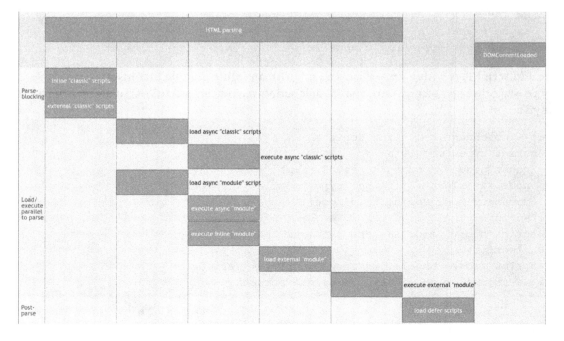

Figure 10.4: Script load/execute order versus browser document parsing

We've now seen how to improve page performance by adapting how JavaScript is loaded and executed. Next, we'll learn how to use resource hints to improve page performance.

Resource hints, per the HTML specification, allow consumers to preemptively complete an operation. They're used as `rel` values on link elements. The values relevant to our use case are `preconnect`, `prefetch`, and `preload`.

`preconnect`'s definition per the HTML standard is as follows:

"`preconnect`: Specifies that the user agent should preemptively connect to the target resource's origin", HTML standard – 4.6.7 link types: `https://html.spec.whatwg.org/#linkTypes`

In summary, `preconnect` allows developers to "tell" browsers to create a connection to an origin, enabling subsequent requests to the origin to occur faster, especially in an HTTP/2 context where more requests can be done in parallel (via multiplexing) and connections are efficiently reused.

For example, we can preconnect to `https://example.com` using the following snippet of code which contains `link` element:

```
<head>
  <link rel="preconnect" href="https://example.com" />
</head>
```

Next, `preload`'s definition per the HTML specification is as follows:

"`preload`: Specifies that the user agent must preemptively fetch and cache the target resource for current navigation according to the potential destination given by the as attribute (and the priority associated with the corresponding destination)." HTML standard – 4.6.7 link types: `https://html.spec.whatwg.org/#linkTypes`

`preload` can be used to load resources before they're detected on the page. This can be especially useful in single-page applications or other highly dynamic JavaScript-driven contexts where resources might not be in the initial returned HTML payload, but we know which resources might be necessary.

Note that `preload` requires a fully qualified resource path (e.g., `https://example.com/assets/resource-1.js`), as opposed to `preconnect`, which uses the origin only. Also, note that `preload` is not designed for use on module scripts; for that, we need `rel="modulepreload"`, which is defined as follows in the HTML standard specification:

"`modulepreload`: Specifies that the user agent must preemptively fetch the module script and store it in the document's module map for later evaluation. Optionally, the module's dependencies can be fetched as well." HTML standard – 4.6.7 link types: `https://html.spec.whatwg.org/#linkTypes`

In our current example, we could request pre-loading of some of our `async` resources ahead of time (before they're "seen" by the browser in the HTML), where our resource loading looks as follows by default. The load order is defined by the order of the script tags in the HTML element and the priority for all resources is `Normal`:

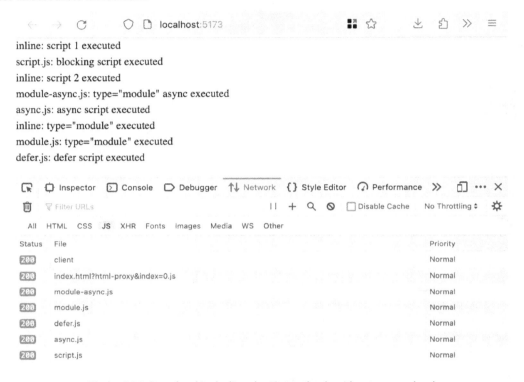

Figure 10.5: Page load including the Network tab without any preload

To illustrate preload, we can add a `preload` link for `async.js` and a `modulepreload` link for `module-async.js` inside the HTML `head` element like in the following snippet:

```
<head>
  <link rel="preload" href="async.js" as="script" />
  <link rel="modulepreload" href="module-async.js"
    as="script" />
</head>
```

If we reload our example page, we'll see that `async.js` and `module-async.js` are now loaded with `Highest` priority, and before the rest of the scripts on the page. Also note that due to the `async` attribute being loaded earlier, the scripts are executed earlier.

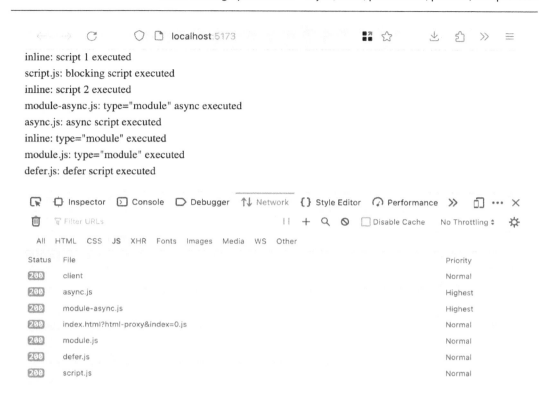

Figure 10.6: Page load including Network tab with async.js having
preload and module-async.js having modulepreload

Finally, prefetch is defined as follows in the HTML specification:

"prefetch: Specifies that the user agent should preemptively fetch and cache the target resource as
it is likely to be required for a follow-up navigation" HTML standard – 4.6.7 link types: https://
html.spec.whatwg.org/#linkTypes

This means that prefetch will not only connect (like preconnect does), but do a full load and
cache cycle. prefetch is useful for when resources will be necessary on the next load as opposed
to for the current page (which is where preload and modulepreload should be used).

We've now seen how to optimize asset loading via the async and defer attributes on script
elements and via preconnect, preload, and prefetch on link elements. Next, we'll look
at how the Next.js Script component's strategy can be used to achieve similar results in a
Next.js application.

Using Next.js Script's strategy option to optimize asset loading

The Next.js `Script` component gives us more control over script loading behavior, allowing us to improve page load performance.

The `strategy` prop allows us to control the loading strategy; it defaults to `afterInteractive`, which will begin loading after some of the Next.js code has run. It can be set to `beforeInteractive`, in which case the script is loaded and executed before all Next.js code. `lazyOnLoad` can be used for lower-priority scripts to delay loading until there's browser idle time.

The final option is experimental; it's the `worker` strategy, which will load and run the script in a web worker.

Per the Next.js docs for the `Script#strategy` option, the following list contains the loading strategies of the script (see the docs: `https://nextjs.org/docs/pages/api-reference/components/script#strategy`).

There are four different strategies that can be used:

- `beforeInteractive`: Load before any Next.js code and before any page hydration occurs
- `afterInteractive` (default): Load early but after some hydration on the page occurs
- `lazyOnload`: Load during browser idle time
- `worker` (experimental): Load in a web worker

One of the benefits of the `Script` component over the `script` native element is that the loading strategy can be used even on inline scripts. For example, say we have a `pages/index.js` page in a Next.js application; we add some `Script` components with two approaches to adding inline scripts. We set the latter `Script` to use `beforeInteractive`, remembering that the default strategy is `afterInteractive`:

```
import React from 'react';
import Script from 'next/script';
export default function Index() {
  return (
    <>
      <h1>Next.js Script Strategy</h1>
      <Script>{`console.log('inline script 1');`}</Script>
      <Script
        strategy="beforeInteractive"
        dangerouslySetInnerHTML={{
          __html: `console.log('inline script 2');`,
        }}
```

```
    ></Script>
  </>
  );
}
```

When we run the Next.js server with `npx next dev` or `npx next build && npx next start`, we see that `inline script 2` is printed in the console before `inline script 1` is; this is the `Script` strategies being applied:

Next.js Script Strategy

inline script 2
inline script 1

Figure 10.7: Second inline Script logging to the console before the first due to the strategy of each Script

We'll now showcase how we can use the loading strategy with external scripts.

Say we have `public/afterInteractive.js`, which contains the following:

```
console.log('afterInteractive.js: loaded');
```

Similarly, `public/beforeInteractive.js` and `public/lazyOnload.js` contain a `console.log` function call with the relevant content, `beforeInteractive.js: loaded` and `lazyOnload.js: loaded` respectively.

We can load them using the following changes to `pages/index.js`; note that we've put them in a rough "reverse" order of loading to showcase the effect of `strategy`:

```
import React from 'react';
import Script from 'next/script';
export default function Index() {
  return (
    <>
      {/* no change to h1 or inline script 1 */}
      <Script src="/lazyOnload.js" strategy="lazyOnload" />
      <Script src="/afterInteractive.js" strategy=\
```

```
        "afterInteractive" />
      <Script src="/beforeInteractive.js" strategy=
        "beforeInteractive" />
      {/* no change to inline script 2 */}
    </>
  );
}
```

When we run the Next.js server with `npx next dev` or `npx next build && npx next start`, we see that `beforeInteractive` is printed on the console before `afterInteractive`, which is printed before `lazyOnLoad`:

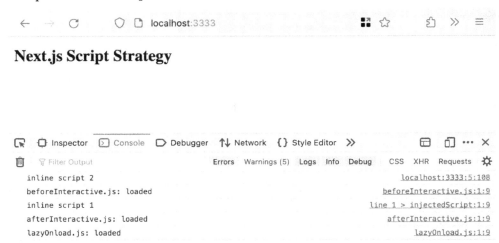

Figure 10.8: Script's logging in order based on strategy

We've now seen how Next.js `Script` and its `strategy` prop allow us to control script asset loading in a Next.js context to achieve additional page load performance. Next, we'll cover how to run scripts in a worker thread.

Loading and running scripts in a worker thread

One of the Next.js `Script` strategy options is `worker`, which loads and runs the script in a web worker. In current Next.js versions, this is achieved via a library called **Partytown** (`https://partytown.builder.io/`). The following is from the Partytown documentation:

"Partytown is a lazy-loaded library to help relocate resource-intensive scripts into a web worker, and off of the main thread. Its goal is to help speed up sites by dedicating the main thread to your code, and offloading third-party scripts to a web worker." Partytown home page – `https://partytown.builder.io/`

To expand on that definition, JavaScript runs in a single-threaded environment in the browser. "Single-threaded" means we only have one entity able to execute compute operations; non-asynchronous work cannot be done in parallel. The main thread in this context is the browser's JavaScript execution thread. When loading and executing compute-heavy scripts, they can starve other scripts of the execution environment. By running said compute-heavy scripts in a web worker, it gets a different JavaScript environment or execution thread, meaning the main thread is freed up to service the rest of the JavaScript execution.

Since `strategy="worker"` for Next.js `Script` is experimental, in order to use it, we need to enable it in `next.config.js` like so:

```
const nextConfig = {
  // no change necessary to other config fields
  experimental: {
    nextScriptWorkers: true,
  },
};
module.exports = nextConfig;
```

When running `npx run dev`, you'll see a warning about the `nextScriptWorkers` experimental feature in the terminal in which you're running the command:

```
▲ Next.js 13.5.4
- Local:        http://localhost:3000
- Experiments (use at your own risk):
    · nextScriptWorkers

✓ Ready in 2.4s
```

To illustrate how we can use `strategy="worker"` powered by Partytown, we can write an `analytics.js` script that will log on, load, and make an API call to `jsonplaceholder` with some information about the page. We store `analytics.js` in `public/analytics.js` to simulate a third-party script being loaded (or more generally, a dependency that cannot be bundled, i.e., one we can't import into our application code):

```
console.log('analytics.js: loaded');
async function trackPageLoad() {
  const responseJson = await fetch(
    'https://jsonplaceholder.typicode.com/posts',
    {
      method: 'POST',
      body: JSON.stringify({
        page: window.location.pathname,
        origin: window.location.origin,
      }),
```

```
      headers: {
        'Content-type': 'application/json; charset=UTF-8',
      },
    },
  ).then((response) => response.json());
  console.log('analytics.js: page load fetch response',
    responseJson);
}
trackPageLoad();
```

We can then create a new `pages/worker.js` file in our Next.js application, which renders a heading and a few Next.js scripts, including `/analytics.js`. The other scripts are to illustrate the load order of the `worker` strategy versus alternative strategy values:

```
import React from 'react';
import Script from 'next/script';
export default function Worker() {
  return (
    <>
      <h1>Next.js Script "worker" experimental
        Strategy</h1>
      <Script src="/analytics.js" strategy="worker" />
      <Script src="/lazyOnload.js" strategy="lazyOnload" />
      <Script src="/afterInteractive.js" strategy=
        "afterInteractive" />
      <Script src="/beforeInteractive.js" strategy=
        "beforeInteractive" />
    </>
  );
}
```

When we load `npx next build && npx next start`, the production server starts, and with the **Console** tab of DevTools open, we can see that `strategy="worker"` loads after all the other strategies. We also see that the `fetch()` call to `jsonplaceholder` completed successfully:

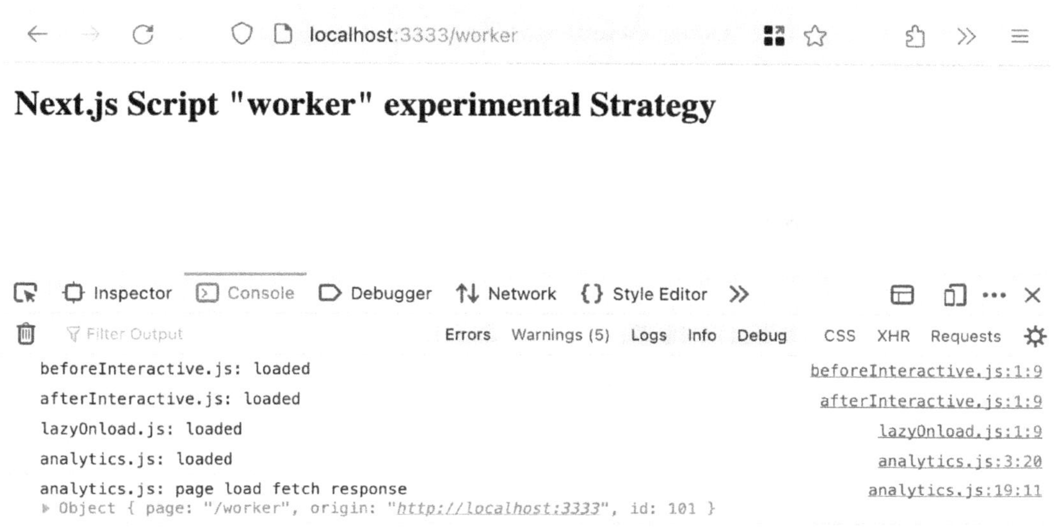

Figure 10.9: worker strategy loading after other strategies and fetch call response logging

Another aspect of loading via the `worker` strategy is that `analytics.js` is not loaded as a script; it's loaded via `fetch`. This can be seen by inspecting the **Network** tab in DevTools, filtering by **XHR** (`XMLHttpRequest`, the precursor to `fetch`) and inspecting the **Initiator** field. Note that the `jsonplaceholder` request appears here (as two requests, an `OPTIONS` request to ensure we can make the cross-origin request followed by the `POST` request).

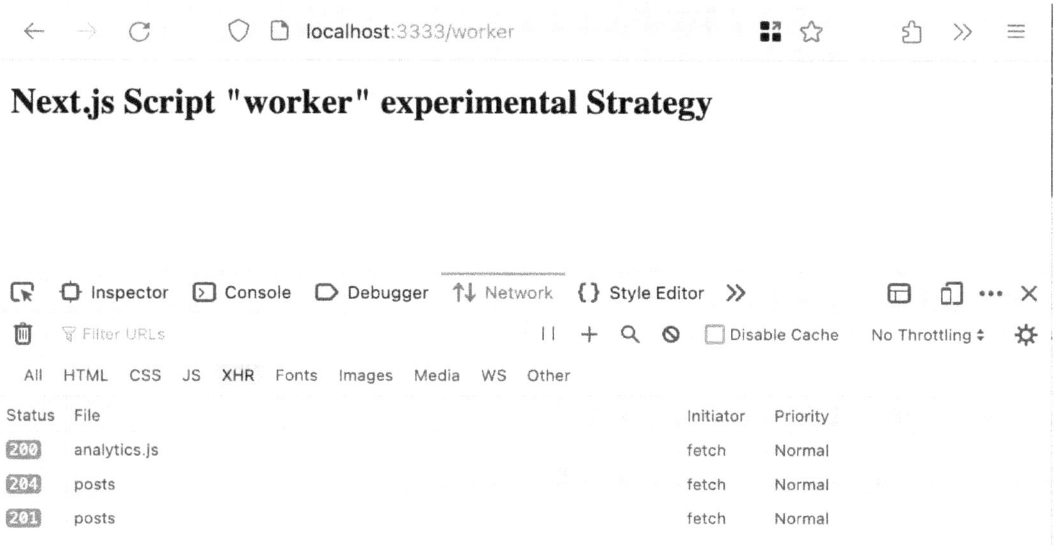

Figure 10.10: analytics.js is loaded via fetch, as are requests to jsonplaceholder

If we dig into the `analytics.js` request further, we'll see that the `Referer` header value (which helps us keep track of the source of the request) is `_next/static/~partytown/partytown-sandbox-sw.html`, which is a Partytown-generated document.

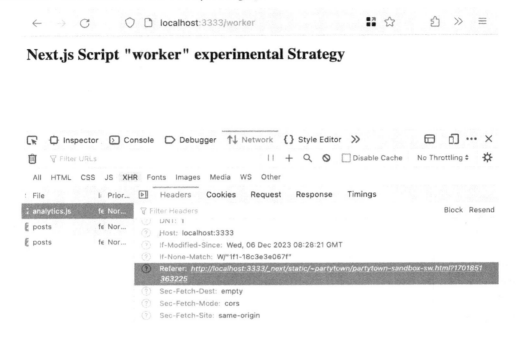

Figure 10.11: analytics.js Referer is the Partytown service worker-generated HTML file

In short, using `strategy="worker"` loads and executes our script in a different JavaScript context to the main window, although Partytown is designed so that it should have a high level of similarity with the origin window.

We've now seen how to use `strategy="worker"` and Partytown to execute scripts off the main thread in a web worker context.

Summary

In this chapter, we've covered techniques to control asset and JavaScript loading more granularly.

In order to control script loading using browser built-in functionality, we can use `async` and `defer` attributes; we covered their effect on module scripts versus classic scripts. We also looked at using the `rel` attribute on a `link` element for resource hints, and what impact `preconnect`, `preload`, `modulepreload`, and `prefetch` have on resource loading.

We can leverage the Next.js `Script` component's `strategy` prop to control script loading and execution beyond `async` and `defer` in the context of a Next.js application.

Finally, we looked at the possibility of running certain scripts off the main JavaScript thread using the Next.js `Script worker` strategy, powered by the Partytown library.

In this final chapter, we covered asset loading strategies and optimizations such as executing code off the main thread.

This brings us to the end of this book. Hopefully, you've achieved a better understanding of design patterns in JavaScript and how to implement them. You will be able to discuss and contrast implementations and the usefulness of language-agnostic patterns that fall into the creational, structural, and behavioral design pattern categories. In addition, you should be confident with JavaScript-specific patterns that will help you scale your applications, reactive view library patterns, rendering strategies, and asynchronous and event-driven programming patterns in JavaScript. Furthermore, you are now familiar with performance and architecture patterns relevant to JavaScript such as micro frontends, lazy-loading, code-splitting, and further asset loading optimizations.

Of course, all these patterns are meant to be used and you will discover new ways to compose them and even notice them in places you didn't expect. The JavaScript and web platform space is ever-evolving, and I hope this book stands you in good stead for using more of its great features.

Index

Packtpub.com

Subscribe to our online digital library for full access to over 7,000 books and videos, as well as industry leading tools to help you plan your personal development and advance your career. For more information, please visit our website.

Why subscribe?

- Spend less time learning and more time coding with practical eBooks and Videos from over 4,000 industry professionals
- Improve your learning with Skill Plans built especially for you
- Get a free eBook or video every month
- Fully searchable for easy access to vital information
- Copy and paste, print, and bookmark content

Did you know that Packt offers eBook versions of every book published, with PDF and ePub files available? You can upgrade to the eBook version at Packtpub.com and as a print book customer, you are entitled to a discount on the eBook copy. Get in touch with us at customercare@packtpub.com for more details.

At www.packtpub.com, you can also read a collection of free technical articles, sign up for a range of free newsletters, and receive exclusive discounts and offers on Packt books and eBooks.

Other Books You May Enjoy

If you enjoyed this book, you may be interested in these other books by Packt:

Mastering JavaScript Functional Programming

Federico Kereki

ISBN: 978-1-80461-013-8

- Understand when to use functional programming versus classic object-oriented programming.
- Use declarative coding instead of imperative coding for clearer, more understandable code.
- Know how to avoid side effects and create more reliable code with closures and immutable data.
- Use recursion to help design and implement more understandable solutions to complex problems.

JavaScript from Beginner to Professional

Laurence Lars Svekis | Maaike van Putten | Rob Percival

ISBN: 978-1-80056-252-3

- Use logic statements to make decisions within your code.
- Save time with JavaScript loops by avoiding writing the same code repeatedly.
- Use JavaScript functions and methods to selectively execute code.
- Connect to HTML5 elements and bring your own web pages to life with interactive content.
- Make your search patterns more effective with regular expressions.
- Explore concurrency and asynchronous programming to process events efficiently and improve performance.

Packt is searching for authors like you

If you're interested in becoming an author for Packt, please visit authors.packtpub.com and apply today. We have worked with thousands of developers and tech professionals, just like you, to help them share their insight with the global tech community. You can make a general application, apply for a specific hot topic that we are recruiting an author for, or submit your own idea.

Hi!

Hugo here, author of *JavaScript Design Patterns,* I really hope you enjoyed reading this book and found it useful for increasing your productivity and efficiency in large JavaScript codebases.

It would really help us (and other potential readers!) if you could leave a review on Amazon sharing your thoughts on JavaScript Design Patterns.

Go to the link below or scan the QR code to leave your review:

https://packt.link/r/1804612278

Your review will help us to understand what's worked well in this book, and what could be improved upon for future editions, so it really is appreciated.

Best wishes,

Hugo Di Francesco

Download a free PDF copy of this book

Thanks for purchasing this book!

Do you like to read on the go but are unable to carry your print books everywhere?

Is your eBook purchase not compatible with the device of your choice?

Don't worry, now with every Packt book you get a DRM-free PDF version of that book at no cost.

Read anywhere, any place, on any device. Search, copy, and paste code from your favorite technical books directly into your application.

The perks don't stop there, you can get exclusive access to discounts, newsletters, and great free content in your inbox daily

Follow these simple steps to get the benefits:

1. Scan the QR code or visit the link below

https://packt.link/free-ebook/978-1-80461-227-9

2. Submit your proof of purchase
3. That's it! We'll send your free PDF and other benefits to your email directly

www.ingramcontent.com/pod-product-compliance
Lightning Source LLC
Chambersburg PA
CBHW080626060326
40690CB00021B/4831